FINDING ZERO

FINDING ZERO

A Mathematician's Odyssey to Uncover
the Origins of Numbers

AMIR D. ACZEL

palgrave
macmillan

First published in 2015 by PALGRAVE MACMILLAN® TRADE
in the United States—a division of St. Martin's Press LLC, 175
Fifth Avenue, New York, NY 10010.

Palgrave® and Macmillan® are registered trademarks in the United
States, the United Kingdom, Europe and other countries.

ISBN: 978-1-137-27984-2

Library of Congress Cataloging-in-Publication Data

Aczel, Amir D., author.
   Finding zero : a mathematician's odyssey to uncover the origins
of numbers / Amir D. Aczel.
      pages   cm
   Includes bibliographical references (pages   ).
   ISBN 978-1-137-27984-2 (hardback)
   1. Numerals—History. 2. Zero (The number)—History. I. Title.
QA141.2.A29   2015
513.5—dc23
                                                      2014024462

Design by Letra Libre

First edition: January 2015

10 9 8 7 6 5 4 3 2 1

Printed in the United States of America.

For Miriam,
who loves science

CONTENTS

ACKNOWLEDGMENTS

I AM EXTREMELY GRATEFUL TO THE ALFRED P. SLOAN Foundation in New York and to Doron Weber, the Foundation's director of the Program for the Public Understanding of Science and Technology, as well as to members of the Foundation's staff, for supporting me in writing this book. I can easily say that without the Sloan Foundation's faith in me, and the research grant it generously provided, this book would not have been written, and the precious stone artifact known as K-127, which bears the earliest zero in our number system, would not have been rediscovered and brought to the attention of the world of science.

Others have helped me on this quest as well. I thank His Excellency Hab Touch, director general of the Cambodian Ministry of Culture and Fine Arts, for his invaluable assistance to me in rediscovering the stele he has since dubbed "Khmer Zero." I thank Chamroeun Chhan, Rotanak Yang, Ty Sokheng, Sathal Khun, Darryl Collins, Takao Hayashi, C. K. Raju, Fred Linton, Jacob Meskin, Marina Ville, W. A. Casselman, Eric Dieu, and especially Andy Brouwer in Phnom Penh for their help.

I am grateful to my agent, Albert Zuckerman of Writers House in New York, for his enthusiasm for this project and his support of this book's publication. Many thanks to my editor at Palgrave Macmillan, Karen Wolny, for believing in this book and for her thoughtful editing, comments, and suggestions, which greatly improved the manuscript. Warm thanks also to Lauren LoPinto for her editing of the book, to Carol McGillivray for her superb editing and insightful comments, to production manager Alan Bradshaw for handling the subtle complexities of producing it, and to copy editor Bill Warhop for his superb editing. Thanks also to designer Rachel Ake, to art director David Baldeosingh Rotstein, and to typesetter Letra Libre for their work in turning the manuscript into a complete book.

Finally, I am extremely grateful to my wife, Debra, for all her suggestions and help, for joining me on parts of the big adventure of searching for the first zero, and for taking some of the photographs in this book.

INTRODUCTION

THE INVENTION OF NUMBERS IS ONE OF THE GREATEST
abstractions the human mind has ever achieved. Virtually everything in our lives is digital, numerical, quantified. But the story of
how and where we got these numbers we so depend on has been
shrouded in mystery. This book tells the personal story of my
lifelong obsession: to find the sources of our numbers. It briefly
traces the known history of the very early Babylonian cuneiform
numbers and the later Greek and Roman letter-numerals, and then
asks the key question: Where do the numbers we use today, the
so-called Hindu-Arabic numerals, come from? In my search, I explored uncharted territory, embarking on a quest for the sources of
these numbers to India, Thailand, Laos, Vietnam, and ultimately
to a jungle location in Cambodia, the site of a lost seventh-century
inscription. On my odyssey I met a host of fascinating characters:
academics in search of truth, jungle trekkers looking for adventure,
surprisingly honest politicians, shameless smugglers, and suspected
archaeological thieves.

FINDING ZERO

1

WHEN I ENTERED FIRST GRADE IN THE LATE 1950s, AT A
private school in Haifa, Israel, called the Hebrew Reali School,
I was asked a question the institution always asked its entering
students. My teacher, Miss Nira, a young and pretty woman who
smiled a lot and wore long, bright-colored dresses, inquired of each
of us six-year-olds what we hoped to learn in school. One child
said "How to make money," and another, "What makes trees and
animals grow," and when my turn came, I answered that I wanted
to learn "Where numbers come from." Miss Nira looked surprised
and paused for a moment, and then without a word turned to the
little girl sitting next to me. I wasn't a precocious child who comes
up with a question the teacher cannot address—I just had a most
unusual childhood. And my answer to the teacher's question was a
direct result of an experience I'd had during that special childhood.

My father was the captain of the SS *Theodor Herzl,* a cruise
ship that sailed the Mediterranean at 21 knots, traveling from its
home port in Haifa to the mythical islands of Corfu and Ibiza and
Malta and—often—to fabulous Monte Carlo. One of my father's
benefits as captain was the right to have his family join him on the

ship whenever he wanted. We took advantage of this privilege very frequently, so that I ended up attending school only part of every year, making up for lost class time with tutors and self-study and, once back at school, by sitting for deferred exams.

As soon as the *Theodor Herzl* would arrive at the charmed principality of Monaco, with its magnificent palace perched on a rock over the Mediterranean, my father would drop anchor and a fast motorboat would ferry passengers and crew ashore. At night, many would head to the great Casino de Monte Carlo by the water's edge near the center of town. This was the undisputed high-class gambling capital of the world. But minors like me were forbidden entry into the famed gaming rooms, where princes and movie stars and celebrities still try to woo Lady Luck. So while the adults from our ship, including my parents, played at the roulette tables, I was entertained by ship's stewards outside this white marble baroque palace surrounded by tall palm trees, bougainvillea, and white and red oleanders. There, my sister, Ilana, and I would run along the paths of this exuberant garden and play hide-and-seek among the fragrant bushes.

Ilana and I often fantasized about what it might be like inside this imposing building we thought we could never enter. Were people dancing? Were they eating fancy meals, as we often did aboard ship? We knew the adults played some kind of game inside—they always talked about it afterward on the ship—but what kind of game was it? We were dying to find out, and whispered about it between us.

And then one day it was my father's personal steward's turn to take care of us on the periphery of the casino. Laci (pronounced "lotzi"), a Hungarian, concocted a scheme to take the girl of three and boy of five right inside the gaming halls. Laci was my favorite steward; the rest of them were boring middle-aged men who took care of us reluctantly (it wasn't really in their job descriptions). They were courteous and polite, but somewhat cold and formal. Whenever Laci took care of us, however, fun things would happen, and often rules of expected behavior were broken. "The boy needs his mother immediately—it's an emergency!" Laci hissed at the stern-faced, hulking, tuxedo-wearing bouncer at the door, and without waiting for an answer ushered us right inside the casino.

I was deathly worried about being kicked out. I knew the casino was taboo for us kids. But to my surprise nothing happened, nobody ran in to chase us away. I was dazzled—on the ornately carpeted floor I saw large, elegant tables covered in green felt, and on each one a checkerboard of numbers in red and black and one very special, round circle of a number alone in green. The atmosphere was heavy with cigar and pipe smoke, and I struggled hard not to cough.

I was excited to see my parents seated at one table. But I knew I dare not disturb them. So I kept very quiet, and didn't move. I feared this dream-come-true might end any minute.

At the head of the table, across from the croupier, was my father, dapper in his black captain's uniform displaying numerous British wartime decorations, and next to him my mother, stunning in a light blue evening dress. They were flanked by a US congressman from a southern state on one side and the famous French-Italian singer Dalida on the other, both VIPs traveling

My father, Captain E. L. Aczel, with the French-Italian singer Dalida, aboard ship off the coast of Monaco, 1957.

with us on this voyage. Other passengers were also seated around this table, and everyone was looking with rapt anticipation at a large black bowl at its center; at the bottom of the bowl was a spinning wheel.

Their attention seemed focused on a little white ball that had been flung into the bowl by a man in a short black coat, a white collared shirt, and a black bow tie. Laci kept inching us closer until we stood right behind my parents. This was so exciting—to be at the heart of this magical activity forbidden to anyone younger than 21. Laci was holding my sister and me, each of us sitting on one of his arms; from our high vantage point above the table I could clearly see what was taking place below.

There was an eerie quiet as the ball twirled around in the bowl. I could hear every time it tapped a metal groove separating numbers at the bottom of the wheel, or when it struck one of the four diamond-shaped metal decorations above the numbers on the inner sides of the bowl, bouncing right back down whenever it did. I could feel the tension and anticipation. My father suddenly turned around when he noticed us, smiled knowingly at Laci, and then turned his attention back to the table.

"Look," Laci explained to me in a whisper, "you see, these are numbers on the table, and notice that every one of them is also on the wheel. Now watch what happens." I sat on his arm and stretched my neck forward—I did not want to miss a thing. The little ball was still bouncing around in the bowl, but more slowly now. Soon it would come to a stop. But where? On which number would it land? Laci told me the ball could choose only one number to land on, since every number was separated from its neighbors by metal dividers. I tried hard to guess where the ball would end up as the wheel slowed down further. I could now make out the separate numbers marked on its bottom.

I was fascinated by these colorful numbers—ornate signs that beckoned me by their mystery, and which as I matured I would understand to represent fundamental abstract concepts that rule our world. I will never forget their shapes on that velvet board. I fell in love with their magic, associating them in my mind with something alluring and forbidden, an unknown pleasure awaiting discovery. The ball finally made one last bounce and came to a stop right on the number seven. Suddenly, a commotion erupted at the table. Across from us an elderly woman in a bright yellow evening gown jumped up from her seat and cried, "Yes!" Every head turned toward her. Some players, perhaps vicariously sharing her big win, congratulated her. Others, maybe envious and upset they had lost, expressed their disappointment.

The croupier forked over a large heap of chips of different colors, the smaller ones round and the bigger ones rectangular with large numbers visible on their faces; I understood these pieces of plastic represented money, each color and shape a different amount. Not knowing much about money at my age, I still could sense by the number and size of these chips, and by the continuing excitement around the table, that this woman had become rich. Laci explained to me that the croupier was giving her many times more than the wager amount since she had bet on a single number. I looked at her, noticed the elation in her face, the happy smile, and heard her rapturous exclamations: "I won! I won!"

Then Laci muttered, as if to himself, "Seven, a prime number." I was curious to know what this meant. Laci always had important things to say, and I knew this whispered exclamation had some meaning.

Later, Laci became my self-appointed math tutor on the ship; he taught me also about prime numbers. They would become a lifelong fascination. One day, having observed Laci teaching me math for some time, my mother asked my father how come he knew so much about the subject. My father told her that Laci had been a brilliant mathematics graduate student at Moscow State University right after the war, but that there had been a big scandal about his research having to do with secret information, perhaps even a suspicion of espionage, and under pressure from the KGB the university asked him to leave. The episode was shrouded in mystery; Laci never talked about it, and nobody knew any details.

But Laci apparently got his revenge on the Soviets, for what happened next was well-known and published in all the newspapers. After he quit his studies, he went to Czechoslovakia to learn to fly military airplanes. Then, in 1948, the Jews of the fledgling state of Israel found themselves attacked by the surrounding Arab states. Laci heard that they badly needed airplanes, so this non-Jew sneaked into the pilot's seat of one of the Czech planes he had been training on, took off, and flew it alone all the way to Israel, handing it over as a gift to the newly founded Israel Air Force.

Then, having nothing else to do in this new land, Laci started working for the shipping company Zim Lines and eventually became my father's steward. Both were Hungarian and shared bonds of common heritage, outlook, and lifestyle. (Incidentally, the *Theodor Herzl* was named after another Hungarian, whose political theory had laid the groundwork for Israel's founding.) My father and Laci were close, and Laci took seriously his role as the captain's steward and was never too far from him. Given his clout aboard ship as the crewmember closest to the seat of power, everyone wanted to be his friend. This was his new life, but Laci never lost his love of mathematics—and he taught me much about it over the years.

"So where do these numbers come from?" I asked him when he put my sister and me to bed on the ship the night of the big casino adventure. "It's a mystery," he answered. "We don't really know." And since Ilana and I were wound up from having been up late in a place we'd always dreamed of entering, he told us part of what he did know as our bedtime story.

"We call these numerals—that's the name for the shapes of the numbers you saw—Arabic," he said, "or sometimes Hindu, or even Hindu-Arabic. But when your father and I were once detained with the rest of the ship's crew in an Arab port city, I spent my time there learning the numerals the Arabs use today." He then opened a drawer and took out a piece of the ship's stationery and drew on it in large print all ten Arabic numerals. "You see," he said, "they look nothing like the numbers we saw on the tables tonight, the numbers we know so well." I looked at these numbers in amazement. I'd never seen such signs. Only the one looked like our 1; the rest of the shapes were totally foreign to us. The five was a small uneven circle, and the zero a dot. I tried to copy the drawn numbers but didn't do too well.

Then Laci took out a deck of cards he had brought with him and placed them face up on a table. I tried to read all their numbers, and little Ilana played with them, turning around the ornately drawn kings, queens, and jacks in black and red. Ilana noticed that they looked the same when upside down. Finally, after playing with these wondrous numbers and shapes for perhaps half an hour, we fell asleep.

The next night, Laci came again to put us to bed. My sister and I shared a cabin next to the captain's, where my parents slept and where we spent our days. "Today let's talk more about the numbers," he said. "So you remember that they don't look very much like the Arabic numerals. Well, guess what: They are not like the Hindi numbers either." Then on a new piece of stationery he drew the Hindi numbers, which some people in India use today, and explained that similar signs are in use in Nepal and Thailand and other Asian countries. "I learned this," he said, "also from experience, when our ship docked at Bombay a few years ago on a world cruise."

I looked at these curious numbers and tried to copy them, and my sister made some drawings. We all laughed at my attempts until we children fell asleep. The following night, I had many questions for Laci about the origin of the numbers: "Where do these numbers really come from? Why don't we use the Arabic numbers? Why do different peoples have their own different numbers? And who invented them all?" I wanted answers right away—I was so curious, so fascinated by them, that I could think of nothing else.

Laci's answer disappointed me a little. All he said was, "Why don't you ask your teacher when you start school next month." This was painful for me to hear. It was late August then, and I knew that, sadly, I would soon have to leave the ship with my mother and sister for at least a few months of school before we could rejoin my father for more cruising. I would miss Laci and the stimulating conversations we had almost every day about ships and ports and cities and sailings—and about numbers.

At school I was often bored. I enjoyed immensely the experiential learning I could pursue aboard ship: finding out firsthand how the world works, even as seen from the still-developing vantage point of a child. I liked being taught about life by my parents and by ship's stewards, and mostly by the dedicated, highly attentive, and instructive Laci. At school, on the other hand, everything seemed rote, divorced from reality, and lacking excitement. There, I simply coasted, doing the minimum and counting the days till we could join my father on the ship. I couldn't wait to be back aboard with my friend and mentor; and my child's intuition told me that he knew much more than he was teaching me.

I had developed an obsession with finding the origin of the ornate numerals I had seen on the gaming table at the casino. I wanted to know where the numerals we use everywhere originally came from, and I was looking forward to Laci telling me more about them—and showing me new things about numbers through our travel experiences. I was barely beginning to understand that there were two related concepts here: numerals and numbers. Numbers were abstract entities. And I felt that there was much more here to be discovered, even if the true depth of the concept of *number* was still beyond my abilities to fully understand. I was mature enough, however, to want to learn how the signs that stand for numbers, the ten numerals we use today, came into being.

THE NEXT INTERESTING SAILING we did with the ship was something special. Instead of the usual pleasure cruises to party islands or opulent casinos (this was before gambling aboard ships had become the norm), the shipping company sent its flagship, captained by my father, on a historical-educational cruise (this, too, was a novelty—it happened before mass "cultural tourism" was born). We sailed first to Piraeus, the port of Athens, and the passengers had an expertly guided tour of the Acropolis and lectures on the birth of Greek democracy, architecture, sculpture, and mathematics.

My father loved the good life—and as ship's captain, he lived it. In every port, the local shipping agent invited him to dinner at the most expensive or most unusual restaurant in town. In Piraeus, the ship's agent was Mr. Papaioanis, a jovial, paunchy Greek originally from the island of Patmos. He invited us all to dinner at a beach restaurant called *To Poseidoncion* (Poseidon's), on one of the backstreets facing the sea. When I think back to that outing so many years ago, I can still smell the fresh shrimp grilling on an open fire and feel the gentle sea wind brush my face; when I close my eyes, I can still see the lights of distant fishing boats gently rolling in the water and hear the waves crashing on the sand. It was a wonderful evening and I hoped it would never end. It was my introduction to Greece and its pleasures. After this wonderful meal, we all took a long walk along the edge of the water, eventually winding our way all the way back to the Port of Piraeus and to our ship, docked by a central pier flanked by two ferries.

The next morning, Laci woke me up early. "When your sister and mother go shopping today, let me take you to see the ancient Greek numbers," he said. "Great!" I answered as I jumped out of bed and started to get dressed. This was going to be an exciting day. I went to my father's large cabin to wait for Laci. He had already prepared my breakfast, and I enjoyed the hot chocolate and the freshly baked, still-steaming croissant made in the ship's ovens.

My father was up on the bridge, and while I was finishing my breakfast—my mother and sister were still asleep—he came down and into the cabin. "You're up early," he said. "Yes," I answered excitedly, "Laci is taking me to Athens to look at ancient Greek numbers!" My father nodded. I wasn't sure that he understood—or appreciated—how important to me my relationship with Laci was: that I would even give up time with my family to go look at some numbers with him.

Laci came, and we went down the gangway to the pier and into a cab that was waiting for us. We drove through the heavy morning traffic of greater Athens, climbing up the densely populated hills between Piraeus and the Greek capital. The pollution was heavy—the Athens area, like Los Angeles, suffers from a thermal inversion pattern that traps gasses and particles—and the bad air made me cough frequently. In about an hour we came to the center of Athens and the cab dropped us off at the Plaka, the city square below the acropolis, full of shops and boutiques and cafes. From there, we climbed up the steep path of the acropolis hill to the Parthenon, the fifth-century BCE temple to the Greek goddess Athena.

We walked slowly up the white stone path. The air was clear and crisp here, and crocuses were blooming; I could smell the fresh scent of the pine trees around us. At the very top, we paid the entrance fee and went into the ruins of the ancient acropolis. We then slowly climbed up the stone stairs to the famous Parthenon. I stopped from time to time to catch my breath and to admire the incredible view of the temple from below as I climbed toward it.

"Beautiful, isn't it?" asked Laci.

"Yes," I said. "It's a very pretty building with these columns."

"These are made of marble," he said. "But you know why you find it so beautiful?" I said I didn't know. "It's the proportion," he said. "The Parthenon follows the ancient Greek proportion called the golden ratio. The ratio of its length to its height is about 1.618, a number that seems to characterize many things we consider beautiful—and it also appears in nature." I was fascinated. He explained that the golden ratio came from a mathematical series of numbers called the Fibonacci sequence, and he showed me how it is obtained, each number being the sum of the two previous ones: 1, 1, 2, 3, 5, 8, 13, 21, 34, and so on. If you divide a number by its predecessor, you get something approximating the golden ratio of 1.618. (For example, the sequence continues with 55 and 89. The ratio of 89 to 55 is 1.61818 . . .) I found this idea enthralling.

We went into the Parthenon and saw some statues, beautiful likenesses of the goddess Athena still with traces of ancient red and gold paint on her face. Laci bent down and pointed to the pedestal of one of the statues and showed me letters carved in the marble. "Look," he said. "This is what I wanted you to see." I saw a letter I did not recognize—it was Greek. "Greek letters were not only used for writing, but also for numbers," Laci said. The letter he was pointing to, he explained, was the Greek delta. It stood for the number four. "This was the fourth statue out of the assemblage that once stood here," he said.

After spending an hour admiring this monument of ancient Greek civilization, we left the temple and began to descend slowly. We stopped to sit down on a flat marble slab that was once part of this columned edifice. From here we could admire the Parthenon. It was hot, and we drank water from plastic bottles Laci had bought. After a few moments, he took out a notebook and showed me how the Greeks used letters as their numerals, and how their arithmetic, using only letters and no zero, worked. Here is what he wrote:

A	B	Γ	Δ	E	F	Z	H	Θ	I	K	Λ	M	N
1	2	3	4	5	6	7	8	9	10	20	30	40	50

Ξ	O	Π	ϙ	P	Σ	T	Y	Φ	X	Ψ	Ω	ϡ
60	70	80	90	100	200	300	400	500	600	700	800	900

The Greek letters used as numbers (including the archaic letters digamma, koppa, and sampi).

The Greek alphabet of that time, the fifth century BCE—the height of classical Greece—included letters that had been extinct since antiquity: *digamma* (F) for 6; *koppa* (ϙ) for 90; and *sampi* (ϡ) for 900. So by the fifth century BCE the Greeks had revived letters in an alphabet they no longer used for writing, just so they would have enough symbols for numbers.

Laci pointed out that the custom of using letters for numbers was rooted in Phoenicia, from which the Hebrew alphabet hails as well. Some Orthodox Jews to this day, he explained, have watches with faces displaying Hebrew letters for the numbers from 1 to 12.

We then went to the museum of Athens, one of the greatest archaeological museums in the world. Among beautiful statues of gods and goddesses I saw several stone inscriptions with numbers represented as letters.

On the way back to the ship, Laci asked the driver to stop in an alley in Piraeus and had me wait in the cab while he went into what looked like a store selling discount electronics. When he came back, he had a small package in his hand. "It's just a small transistor radio," he said to me. These were the new popular prod- ucts of the time, the early 1960s, and people were going crazy over them. You would see a person walking down the street listening to a transistor radio as often as today you might see someone talking on a cellular phone. "Just a present for someone," he said. I thought nothing more about it at the time.

After leaving Piraeus that evening, the *Theodor Herzl* sailed to Naples, and the passengers spent the day visiting nearby Pompeii. Next, there was a stop at Civitavecchia, the modern-day port of Rome (in antiquity Rome's port was Ostia). The passengers had a tour of the city, with an in-depth study of Rome and the history of its empire. For a boy interested in the history of numbers, this was a cruise to remember.

In Pompeii, Laci and I traced the Roman numerals—all Latin letters—that were used for house numbers in this ancient town. Its ruins were in a remarkable state of preservation as they had been covered by volcanic ash for almost two millennia after the catastrophic eruption of Vesuvius in 79 CE. At its museum we saw more numbers written using letters, and it was fun to read them and even try basic arithmetic.

Rome was a feast for a budding number-hunter. Roman numerals were everywhere. Most interesting were the milestones the Romans had placed along their straight-as-a-ruler roads, and I slowly became adept at reading the distances on these ancient markers, which I saw in museums and on the most famous of Roman roads, the Via Appia Antica.

Laci explained to me how the Romans devised their number system. He drew for me all the Roman numerals: I is one, V is five, X is ten, L is 50, C is 100, D is 500, and M is 1,000. He

explained that this method made it necessary for the written numbers to grow, and grow, and grow . . . He showed me how a Roman
might have had to multiply XVIII by LXXXII, and ultimately get
the answer MCDLXXVI. For us today, this operation is simply
18 × 82 = 1,476, and we can do it quickly and efficiently. Laci
challenged me to perform such a calculation in the Greco-Roman
system and made me construct its multiplication table; it was so
huge and complicated that it took me a week to do. Amazingly,
he said, this inefficient numerical system remained in wide use in
the West until the thirteenth century, when it was replaced by the
numbers we know today.[1]

I learned more about numbers from this lover of mathematics than I ever did at school, and I was grateful to him. But the
mystery of where the ten numerals we use today originally came
from continued to haunt me. At the same time, as I matured
through school and pursued adventures to discover numerals while
traveling the Mediterranean aboard my father's ship, I began to
understand the abstract—and even more mysterious—concept
of a number. I realized that 3, for example, stood for the *idea* of
"threeness"—something that was shared by all things in the universe that were three. All of them were to be described, in their
quality of threeness, by the unique symbol 3. Equally, 5 stood
for the quality of "fiveness" shared by everything that was five in
number. This fascinating discovery made me even more eager to
find where the numerals came from since they actually stood for
something even deeper and more alluring than I could have ever
imagined as a young child. I wanted to dedicate my life to traveling
the world in search of an answer to the origin of numbers. Who

invented these wondrous ten numerals? I asked myself this question all the time, and also, Who ever came up with the amazing idea that a concept of "threeness" or "seventeenness" or "three-hundred-and-five-ness" could be captured simply by a combination of ten signs arranged in certain ways?

2

IN 1972, AFTER HIGH SCHOOL AND THE OBLIGATORY
service in the Israeli army, I was accepted as an undergraduate
in mathematics at the University of California at Berkeley. I now
made one more voyage on my father's ship. By then, all the passenger ships had been sold because Zim Lines had lost too much
money due to poor management of the company's cruise division,
called Zim Passenger Lines, and my father was now the captain of
a small, slow, old cargo ship, the *M.V. Yaffo,* which sailed between
the Mediterranean and the Americas. So I hitched a ride aboard
my father's ship, embarking in Haifa in late July for the long trip
that would ultimately bring me to Berkeley.

A cargo ship is as different from a cruise ship as a truck is
from a limousine. Like a working truck, a cargo ship can be dirty
and dusty, but its cabins are roomy; with no passengers, there's
no need to cram many people into limited cabin space, and the
crew is therefore more comfortable. But the flipside is that there
is nothing to do: no cocktail parties or bars or ballrooms, and no
exciting social gatherings. A voyage on a cargo ship can be lonely.

But Laci was still my father's steward, and I was very fond of him. We often talked about mathematics during this trip.

As an adult, I now understood that the mystery that had held my attention since early childhood was really two mysteries. One was where the first numerals originated: In which part of the world, and when, did people first invent the nine numerals, and zero, that with time evolved into the numbers that rule our world? And the second was a deeper conundrum, which I was now sophisticated enough to discern: How did humans abstract the concept of number? How did the idea of a number originate, and how did it develop and mature through history, to bring us to the digitally dominated society in which we live today?

Laci and I spent many hours together discussing this latter idea as the ship slowly made its way across the ocean. It was a far deeper discussion than I had ever had with him as a child, certainly, and he brought to it his full intellectual abilities as a one-time doctoral student of pure mathematics at the top Russian university, where mathematics has always been one of the most important fields of study. It was a pleasure to sit with Laci on deck chairs discussing mathematical concepts I was only beginning to understand and be fascinated by—building on the earlier ideas he had taught me as a child: numerals, numbers, prime numbers, and the mysterious Fibonacci sequence.

HAVING FINALLY CROSSED THE ATLANTIC, we docked for a few days in Halifax, Nova Scotia, and then continued to New York City; Charleston, South Carolina; and finally Miami, where I disembarked before the ship continued on to the Caribbean and

South America. Before I left ship to fly to San Francisco and start my studies, Laci said one last thing to me in parting: "Remember how when you were little we used to talk about where the numerals came from? Maybe you'll find out. I once read in a science magazine that a French archaeologist may have found something about the numbers in Asia decades ago—something important relating to the zero. But I don't remember any of the details."

Laci's parting words intrigued me, but I had no opportunity to pursue this research further. At Berkeley, I had a full plate of math courses, challenging but often enjoyable, and I had to worry about grades and exams and learning to become a mathematician.

However, through my courses—mostly mathematics, but also anthropology, sociology, and philosophy—I learned a fair amount also about numbers and their development.

Numbers, as a concept, are much older than we might think. In 1960, a Belgian explorer named Jean de Heinzelin de Braucourt was surveying the region of Ishango, at the border of present-day Uganda and Congo (then the Belgian Congo), when he discovered a strange-looking bone: a baboon's fibula bearing what looked like numerous tally marks. Analysis later concluded that these markings might evidence very early counting. The bone has three sets of identical notches, adding, respectively, to the totals 60, 48, and 60. The markings are grouped in several sets containing 5, 7, 9, 11, or 13 tally marks each. This bone was scientifically dated to about 20,000 years ago—the Paleolithic era—when humans lived in hunter-gatherer groups. The Ishango bone provides some of the earliest known evidence of a form of

The Ishango bone: a baboon fibula, about
20,000 years old, bearing notches believed
to represent early evidence of counting by our
species.

counting by humans who lived in Africa so long ago, and it is now displayed at the Royal Belgian Institute of Natural Sciences in Brussels.

What does the Ishango bone represent? It seems that in prehistoric times, early humans roaming the bush in Africa used bones of dead animals as a way of drawing the first *one-to-one correspondence* between the number of animals they were able to hunt and notches they made on a piece of bone. This was not quite counting, but it was close. Anyone could see that a bone that had more such tally marks implied that its owner had hunted more animals—perhaps without really knowing what the actual *number* was. Of course all this is merely a hypothesis, but it is a likely one.

The Ishango bone is certainly the best early example of pre-counting. But there is some evidence that European humans may also have used some kind of pre-counting: In addition to the Ishango bone, several animal bones with markings that are likely tally counts have been discovered in Europe and also dated to the Paleolithic.[1]

A later piece of evidence for something resembling counting comes from the mysterious Neolithic stone arrangements, believed to be about 6,000 years old, at Carnac on the coast of Brittany in France. Interestingly, the megaliths are found in groups, and the groups are often comprised of a prime number of megaliths: 7, 11, 13, and 17. Was this by chance, or does it represent a form of counting? Does it, perhaps, represent an even deeper understanding of numbers? We don't know. Carnac is a mystery that archaeology has never been able to explain, despite attempts over many decades. Nobody knows why so many very heavy stones were placed in rows. Perhaps a connection exists with Stonehenge, where similar stones were placed in circles around the same time.

But Ishango and Carnac do not present what we consider numbers. In everyday life, numbers began with simple representations of a quantity by people using their fingers. Aristotle wrote two-and-a-half millennia ago, "Or is it because men were born with ten fingers and so, because they possess the equivalent of pebbles to the number of their fingers, come to use this number for counting everything else as well?"[2] And since we also have ten toes, early societies used these as well to count beyond ten. Remnants of such a base-20 number system that existed long ago can still be seen in French, through words such as *quatre-vingt* (four twenties) for 80.

Clearly, the form of counting we use evolved from an accident of nature: our having five fingers on each hand and five toes on each foot. Indeed, in some languages, such as Old Khmer, five was used as a point to anchor the other numbers: After four comes five, then five-and-one, five-and-two, and so on to ten, which becomes the next anchor. When we look at the early European numbers, the Roman numerals, we see the same trend: The Roman numbers IV, V, VI, VII, VIII, are all anchored at five (V), and only after eight do we come to measure them in relation to ten: IX, X, XI, XII, XIII. So the use of five and ten as key numbers evolved in different parts of the world.

In ancient India, the number 10 was the anchor. Decimal numbers were evident very early on there, as early as the sixth century BCE as observed in some inscriptions. And once they learned the powers of ten—simply by using ten fingers, and then using ten fingers ten times, as in ten people, one for each finger, each holding up his or her ten fingers—the Indians of antiquity understood that this process can go on forever. Ten times ten people holding their fingers up was ten to the third power, and ten times ten times ten people with ten fingers each was ten to the fourth power, and so on without a limit.

During the Han Dynasty in China (206 BCE to 220 CE), a mathematical work titled *Nine Chapters on the Mathematical Art* appeared, employing both positive numbers, colored red, and negative ones, colored black. And in Egypt of the third century, a leading Greek mathematician named Diophantus obtained negative answers to some of his equations but immediately dismissed them as unrealistic. So the idea for negative numbers is quite old, but people did not understand such numbers. The double-entry bookkeeping system used in accounting today was developed in Europe in the thirteenth century in part to avoid using negative numbers. To define negative numbers requires the concept of *zero*.

Negative numbers are, in a sense, a reflection across zero of the positive numbers. You can see this if you draw the number line, starting at zero and going to the right to 1, 2, 3, and so on; and then extending the numbers to the left of zero to –1, –2, –3, and onward. The number –1 is a reflection of the number +1 through a mirror placed at zero. But the zero plays many other important roles in mathematics and its applications. And some crucial equations in physics, biology, engineering, economics, and other fields use zero as the key element (Maxwell's equations in physics are a good example).

THE NUMBER LINE

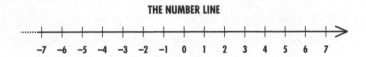

The ancient Babylonians developed a cumbersome number system based on 60 rather than 10, and without a real zero. In this system, ambiguities arose because of the lack of a zero, and they could only be resolved from the context. As a modern example, if someone told you that something costs six-ninety-five, you would understand it to be $6.95 if you were buying a magazine and $695 if you were purchasing an airplane ticket. With their cumbersome number system, the Babylonians were forced to make such assumptions all the time. But interestingly, vestiges of their ancient system still exist today: We have 60 seconds in a minute, 60 minutes in an hour, and the circle has 360 (6 × 60) degrees. All in all, however, the Babylonian system would be as inadequate today as doing calculations with one's fingers and toes.

An interesting question arises: Why use a base of 60? We have ten fingers, so base 10 makes sense, and if you insist on also counting toes, a base 20 may be useful. But 60? In 1927, the prominent Austrian American historian of science Otto Neugebauer suggested that the choice of the large base of 60 was made to address an important practical problem in using numbers in Babylonia. Often, fractions of a whole, such as ½, ⅓, ¾, and ⅔, were required as measures: Perhaps someone wanted to buy half a loaf of bread, or a third of a wheel of cheese, or two-thirds of a shepherd's pie. How could the numbers ½, ⅓, ¾, and ⅔—the most commonly used fractions—be reconciled with a natural system using ten numbers abstracted from fingers? Neugebauer's answer was that 60 is a good solution since this number is divisible by 2, 3, 4, and 10, and for this reason it was chosen as the base for the entire system. Another hypothesis is that the Babylonians knew five planets (Mercury, Venus, Mars, Saturn, and Jupiter) and that they chose their base, for cosmological reasons, to be the product of this number and the 12 (lunar) months of the year.[3]

I already knew something about the Greco-Roman number system from visiting Greece and Rome. This system, too, lacked a zero, and with it the ability of the numbers to cycle so that the same signs could be used over and over again to mean different things. The Greco-Roman system, like the Babylonian and Egyptian, had to fade away, remaining only as an elegant way of commemorating official dates or representing time on clock and watch faces.

Then, in the thirteenth century, a system of numbers consist-
ing of nine numerals and a round zero appeared in Europe. This
innovation became popular and within a few decades took hold in
all segments of educated society. Merchants, bankers, engineers,
and mathematicians found that it improved their lives because
they could make quicker calculations with fewer errors.

It is believed that Leonardo of Pisa (ca. 1170–1250), better known as Fibonacci (of the famous Fibonacci sequence), was
the one to bring the Hindu-Arabic numerals to Europe. He did
it through his book, *Liber Abaci* (the book of the abacus), published in 1202 and circulated widely throughout the continent.
This mathematical volume described the nine Indian figures—the
digits 1 through 9—and a symbol, 0, for what Fibonacci called
zephirum, meaning zero. The root of the Latin zephirum has been
traced to the Arabic word for zero, *sifr.* So a linguistic connection is
found here from the Arab zero to the new European one. And the
author clearly refers to the nine digits as "Indian." We thus have in
this one source an indication both for an Indian and for an Arabic
origin for our modern numbers.

This number system introduced into Europe in the late Middle Ages was far superior to the Roman one used until then. It allowed an immense economy of notation so that the same digit, for
example 4, can be used to convey itself or forty (40) when followed
by a zero, or four hundred and four when written as 404, or four
thousand when written as a 4 followed by three zeros (4000). The
power of the Arabic, or Hindu, or Hindu-Arabic number system
is incomparable as it allows us to represent numbers efficiently and

compactly, enabling us to perform complicated arithmetic calculations that could not have been easily done before.

But the real origin of our amazing number system based on nine digits plus a place-holding zero remained a mystery. The nine numerals had been conjectured to originate in India, as Fibonacci had implied, but there was no unambiguous scholarly proof of this belief. And the zero: Was it Arabic, or Indian, or did it come from some other place? I still didn't know.

3

I EVENTUALLY BUILT A CAREER AS A MATHEMATICIAN
and statistician. For several years, I was a professor of mathematics
at the University of Alaska in Juneau, and there, in 1984, Debra
and I got married just below the Mendenhall Glacier, surrounded
by Douglas fir trees, with occasional brown bears coming in from
the forest to feed on the salmon making their way up the Mendenhall River. We had met at the university a few months earlier,
when I helped Debra with a statistical problem about how well the
university attracted and retained students from a diverse population of Alaska natives and Americans who had moved there from
the Lower 48, and we fell in love.

Some years after our wedding by the glacier, we moved to
Boston and I started teaching at Bentley University. Debra took
a position running a program at MIT, our daughter Miriam was
born, and I wrote a number of popular books on the history of
mathematics and science.

In 2008, I got a call from my friend Dr. Andres Roemer, a
Mexican intellectual who had studied policy at Harvard, got his
PhD at Berkeley, and was the host of popular television shows

produced in Mexico and aired throughout the Spanish-speaking world. Andres invited me to speak about probability theory at an international conference he was inaugurating. After the conference, Debra and I went to see the National Museum of Anthropology in Mexico City. Unexpectedly, this visit to a museum rekindled my childhood interest in the origins of numbers.

In the main hall of the museum, on the wall facing the entering visitor, we were greeted by the stunning Aztec Stone of the Sun. A circular stone 12 feet in diameter and weighing more than 24 tons, this artifact, attached to the museum's wall, bears a central face, believed to represent the Aztec god of the sun, Tonatiuh. Around it are markings and designs that have never been

The enigmatic Aztec Stone of the Sun at Mexico's National Museum of Anthropology.

deciphered. This might have served as an ancient calendar. This curious archaeological find reminded me of something I had seen decades earlier with Laci on our visit to Athens.

Just below the Acropolis, at the periphery of the Plaka, there stands an ancient Greek tower dated to the second century BCE. This tower is octagonal in shape to represent the eight winds that navigators have recognized since antiquity—each wind blowing from one of the compass directions: north, northeast, east, southeast, south, southwest, west, and northwest. Could there be a connection? I wondered: Could the Aztec Stone of the Sun also represent the eight cardinal directions, just as the Tower of the Winds had?

Debra and I marveled at this ancient Aztec artifact with its perfect geometrical design and intricate markings. We wondered what it truly represented and what its purpose might have been. We talked quietly about it for almost half an hour, standing mesmerized by this mysterious stone carved by mathematically inclined people centuries ago. A nearby display explained that the Aztec Stone was likely made in the fifteenth century and had been discovered right in the center of Mexico City.

Some years earlier, I was surprised to learn from an anthropologist friend that much scientific research is actually done inside museums—not just in the field, as one might expect. Museums present collections of artifacts that have been cleaned and prepared for display and are usually shown within the context of similar items that are related to them by date or type or location of discovery, or all of these. This practice facilitates analysis by experts as much as admiration by the general public. Recently, many great museums around the world have begun enhancing their exhibits with video presentations about topics related to their displays, which serve an important public-education need.

And when Debra and I went upstairs from the Aztec Stone, we serendipitously found an ongoing video presentation on Mesoamerican mathematics. I was fascinated to learn that two millennia ago, the Mayans had devised a sophisticated calendar using glyphs for numbers—including a zero. The Maya numerals go back to 37 BCE, and they are simple to write. The numbers 1 through 4 are dots, 5 is a bar, and 10 is two bars one on top of the other; and for zero there is a crescent-moon glyph.

In fact, the Maya invented four kinds of calendars. One was the Long Count calendar, which represented days from a starting point that corresponds, in our modern calendar, to August 11, 3114 BCE—the date of the creation of the universe, according to Mayan mythology. The counting of days from creation used a mixed base-20 and base-18 number system. One of the digits in this calendar would reset to zero after reaching 18; otherwise the calendar used the vigesimal (base 20) numbers. Surprisingly, the ancient Maya, who inhabited the Yucatan Peninsula in Mexico and parts of Central America, apparently understood the concept of zero as early as the first century BCE.

The Maya also had a Short Count, a cyclical calendar with 260 days (20 times 13 days), which was their sacred calendar. At the end of the cycle, monuments were erected to commemorate the fulfillment of the period. A third Mayan calendar had 360 days—close to the usual solar year of 365.24 days. It was for the construction of this calendar that the usual base-20 Mayan number system was adjusted to also employ a base of 18, because 360 equals 20 times 18. Had they used only multiples of 20, their year would have been forced to have 400 (20 × 20) days.[1]

Yet a fourth calendar used by the Maya was based on cycles of the planet Venus. The Mayans were astute observers of the sky and had long ago noticed that Venus would rise with the sun (we call this a heliacal rising) every 584 days—so the calendar based on Venus reset itself to zero after 584 days. The Mayan calendars, and the predominantly vigesimal Mayan number system with zero, are some of the most intriguing discoveries in the history of science. In 2012 there was worldwide panic in some circles of society fearing the end of the world because one of the Mayan calendars reset itself to zero. Of course, nothing happened; our planet continued to revolve around the sun, and this fear turned out to be as unfounded as the similar Y2K worry of a dozen years earlier.

But the Mayan system was isolated from the rest of the world, and it used glyphs—written or carved symbolic signs—that were not suitable for economy of notation. Their signs grew in number as the numbers they represented got larger, in a manner similar to that of the Roman system. The zero was not a perfect positional element as in our numbers, and the base changed, depending on need, from 20 to 18. Georges Ifrah calls the Mayan numbers "a failed system"—it was not one that survived the test of time.[2] Encountering it, however, rekindled my passion to search for the origins of our versatile base-10 numbers and the first zero in the East—the ten numerals that became the basis for the all-powerful system that controls our modern world.

4

OVER THE NEXT YEAR—ENERGIZED BY LEARNING ABOUT
the Mayan numbers—I worked hard to solve the great mystery
that had intrigued me my entire life: Where do numbers—*our*
numbers, the nine familiar digits, 1, 2, 3, 4, 5, 6, 7, 8, 9, plus the
all-important 0—come from?

I knew from Laci and from my university courses and extensive reading and research I had done that the nine digits we use
today were believed to have originated in India. I also knew that at
some point in the past, the Indians learned to use a place-holding
zero. But there were no absolutely known facts about the origins
of the numerals and the birthplace of the zero and no details. Was
all this true? Books and articles all pointed me in one direction:
East. My experiences with Laci on the ship gave me a deep, lifelong
desire to find things out for myself, to see the actual evidence, to
witness history.

So I began to plan a trip to India, hoping to find some answers
there. I spent much time learning about Hinduism, Buddhism,
and Jainism. I read books about the East and the cultures of Asia,
about customs, philosophy, and mathematics. I sensed that Eastern

religions were the key to knowledge about Asian societies, and I felt that perhaps the origin of the numerals was embedded in these religious traditions.

And here is what I learned about these fascinating religions, which until then were almost completely foreign to me. Brahmanism, more commonly known as the precursor of modern Hinduism, began in India and has three main gods: Brahma, Vishnu, and Shiva. Each god has a feminine aspect called his Shakti, or consort. Parvati is Shiva's Shakti; she is also called Uma or Durga. Lakshmi is Vishnu's Shakti. She emerges radiant from the Sea of Milk, seated on a floating lotus and holding a budding lotus flower in each hand. She is the bearer of good fortune. Brahma is the creator of the worlds, but he is born from a lotus flower on Vishnu's stomach once Lakshmi wakes him up by massaging his legs as he lies in eternal slumber on top of the sea serpent, Ananta—which means infinity. Here we already see that at the very point these gods were conceived, the key mathematical idea of infinity makes its debut in the form of an infinite quantity or extent, as embodied by Ananta, and in the form of an infinite past: eternity until Vishnu is awakened.

Now, Vishnu is the maintainer of the worlds, and Shiva is their destroyer. In keeping with this powerful image, Shiva is often portrayed as holding a trident; at times he is represented by a stylized phallus. For a dangerous god he seems rather benevolent: His chief interest, as evident from much of the art in which he appears, seems to be sex.

Vishnu has four (or sometimes eight) arms. His four arms carry symbols representing the four elements of the cosmos: earth, wind, fire, and water. These are surprisingly similar to the Greek elements of the universe, which are also earth, wind, fire, and water, but which include a fifth essence called just that: quintessence, from which we get the word *quintessential.*

Shiva is almost always portrayed as having a third, vertical eye on his forehead. And Brahma has four faces, one for each of the four cardinal directions: north, south, east, and west. These gods seem to capture much symbolism about life and nature. Together, the three gods form the Trimurti—reminiscent of the trinity. And in fact Vishnu and Shiva are sometimes depicted as a single amalgamation: a statue with four arms *and* a third eye, representing a god named Harihara (Vishnu is Hari, and Shiva is Hara). Together, the two gods, or three when Brahma is included, are seen as representations of a single supreme being.

My research in preparation for the voyage to India gradually convinced me that there must have been something about the *philosophy* of the East—perhaps Buddhism, Jainism, or Hinduism, or maybe some ideas from all three religions together—that made the Eastern mind more amenable to completing the number system on *both* of its extremes: adding a zero at one end and infinity on the other.

We know that in Europe, before the arrival of our numbers, there was no zero. Numbers could be added, subtracted, multiplied, and divided, but no one thought of a zero. If you subtracted 5 from 5, for example, then there was nothing—but not the something we now call a zero. The computation simply ended at that point. Equally, Europeans never contemplated extremely large numbers and infinity—as I knew the Jains of India had definitely done—with the exception of religious philosophizing about God's infinite qualities. For example, in *The City of God*, St. Augustine writes: "Of God's eternal and unchangeable knowledge and will, whereby all He has made pleased Him in the eternal design as well as in the actual result."[1] This is an oblique reference to infinity, to eternal time, but the concept is not developed any further. Similar references to God's infinite qualities and working in infinite settings of time and space are found in Jewish theology, especially Kabbalah, but there, too, these notions are vague and not yet well formed.

I became convinced that it must have taken an Eastern mind, perhaps using a unique and very different kind of logic, to invent both the idea of a zero and the concept of infinity. While I could not know it at the time, in the East my hunch would turn out to be truer than I initially expected. I was eager to travel to Asia and hoped to find there and see with my own eyes in an ancient manuscript or on a carved stone—some of the first numerals recorded in our distant past. The thought of actually finding archaeological proof for our mathematical origins filled me with indescribable excitement. India would be the first place to look.

I LANDED IN FOGGY DELHI at 2 a.m. on January 10, 2011, in the dead of winter. As I learned the hard way, while the south of India is tropical, Delhi, in the north, can be awfully cold. It was damp and freezing as I entered the terminal. I was shivering, deathly tired, and somewhat disoriented from a very long flight from America. I had with me a small suitcase, a couple of books about Indian mathematics, and a notebook with the name of a person I hoped would help me address the mystery.

Two years earlier at an international conference on the history of science in Sydney, Australia, I had met Professor C. K. Raju, perhaps the most unusual academic I have ever encountered. At that meeting, Raju had given a talk that elicited shock, skepticism, and even jeers. He claimed that mathematics was born in India and that much of what we in the West attribute to the mathematicians of ancient Greece had, in fact, been achieved earlier in India. He didn't offer much by way of definitive historical proof, but he made up for it with his passion, his own conviction that he was right, and his considerable personal charm.

There was something about his talk—and personality—that made me believe that maybe, just maybe, this man was not talking nonsense, as it sounded to most of us at the time. I knew there were many documents on Indian history that were unknown in the West (ancient Indian documents number in the millions), and I felt there was a chance that at least some of the mathematical derivations and results Raju was claiming for India might indeed have first been discovered in the subcontinent and later transmitted to Greece or other places.

For one thing, Pythagoras himself might have traveled to India in the fifth century BCE, as we know he had visited Egypt and Phoenicia. I had spent some time chatting with Raju and we remained in touch. Now that I had come to the East, I had arranged to meet him in the lobby of the opulent Oberoi Hotel in the heart of New Delhi.

Everything about the East seemed ruled by a bizarre kind of logic. We were supposed to meet at the hotel lobby at 2 p.m.; I sat there for two hours drinking a succession of cups of Assam tea and was about to give up at 10 minutes to 4, and then Raju appeared. He didn't apologize or explain his lateness—it seemed natural to him, I suppose, that 2 o'clock or 4 o'clock were pretty much the same thing. After some small talk about our mutual acquaintances from the Sydney conference, he started on a long monologue about science and mathematics, claiming for India many of the world's discoveries of the past three millennia, from theorems we usually attribute to the Greek mathematician Euclid to facts about relativity for which we normally credit Einstein. Then, referring to my project, he said, "Correcting Western bias in the history of science would be something you should definitely do." He then opened a book and showed me a verse in it, surprising me with the bizarre logic of the East:

Anything is either true,
Or not true,
Or both true and not true,
Or neither true nor not true.
This is Lord Buddha's teaching.[2]

Upon reading these words, I exclaimed, "God, what strange logic!" Raju laughed and explained that it came from the prominent second-century CE Buddhist philosopher and teacher Nagarjuna. He looked at me and said, "Spend some time in the East and you will understand it."

I stared long and hard at Nagarjuna's statement, and considered. True, not true, both true and untrue, neither true nor
untrue?—What is this set of options? They made no sense at all.
This certainly was a bizarre way of thinking about reality. Raju
quickly explained, "The key to everything here is called *Shunya*."
He smiled widely, and when he saw that I did not understand what
Shunya meant, he continued. "Shunya means zero in our language.
But it also relates to the Buddhist philosophical concept of the
void, which is called *Shunyata*. You see, zero, the number, and the
Buddhist emptiness—the goal of meditation and an ideal striven
for on the road to Nirvana, or enlightenment—are one and the
same. Emptiness is a deep philosophical concept, and from it we
get our zero."

Professor C. K. Raju in Shimla, the foothills of the Himalayas, India.

Then Raju picked up his worn leather bag full of papers, shook my hand a few times, and said, "But you will figure it all out." He flashed his big, toothy smile again and disappeared. He had to rush, he had explained; he was leaving for Malaysia, where he had a visiting professorship at the university. The Malaysians, apparently, were interested in work that would bring scientific recognition to Asia and were willing to pay for it; Raju was regularly presenting the research he was producing in Malaysia at international conferences in Germany.

The Germans, too, were involved in projects on the history of Asian science, and for them Raju was a gold mine, someone who prodigiously produced papers on this topic. I was glad for the idea about Shunya, the zero, being derived from the Buddhist void, Shunyata. I also had a vague feeling that the different kind of logic—true, untrue, both true and untrue, and neither true nor untrue—somehow was related to the concept of the void, and hence to the Eastern zero.

5

I BEGAN MY SEARCH IN SOUTH ASIA WORKING BACK-
ward in time, starting with the tenth century CE. The day after my
memorable meeting with Raju, I took a yellow-green tuk-tuk—the
ubiquitous three-wheeled, motorcycle-powered vehicle that can
navigate even the narrowest streets of the most crowded of Asian
cities—through Delhi's early-morning mist to the airport. There
I boarded a Kingfisher Airlines flight to Khajuraho, a complex of
Hindu and Jain temples situated in what was once a dense forest in
the state of Madhya Pradesh. After two hours of flight, the Kingfisher stopped at the holy city of Varanasi on the Ganges, where
we stayed on the ground for about 20 minutes, and then took off
again for Khajuraho, less than an hour away. As we were about
to land, I glimpsed stone temples below in the remains of a lush
tropical jungle.

A few decades ago, a Japanese mathematician named Takao
Hayashi had photographed some mysterious numbers here, but he
had no information on the name of the temple where the ancient
inscription could be found. I had to discover it for myself.

Khajuraho has a tiny airstrip, and a hut serves as a terminal. I got a cab, a rickety old Ford with no upholstery or much of an interior, smelling of sweat and rotted vegetables. I took it up the straight, dusty road through dried-up, hardscrabble fields to one of just a handful of hotels, a Best Western—the nicest place in this tiny town, it turned out. The young man at the desk was not as interested in assigning me my room as in selling me tourist trinkets in his side business. I turned him down and asked for my room key—I would have no time for shopping, I explained. After checking in, I left the hotel and in about 15 minutes of walking on deserted roads managed to find my way to the temples. A man with apparent signs of leprosy sat on the ground outside the entrance to the fenced compound. I paid the admission fee and entered. Most tourists come here to gawk at the graphic erotic statues and friezes that adorn these unusual temples.

In 1838, a British military officer, Captain T. S. Burt, was exploring the jungles of Madhya Pradesh some 400 miles southeast of New Delhi with his company of Bengal Engineers when he and his men came upon a group of ancient temples that had been reclaimed by the jungle. What they saw stunned Burt. In his logbook, he noted that these temples were among the finest he had ever seen, but he was also at a loss on how to describe the nature of the erotic art he saw at Khajuraho. About 10 percent of all the magnificent, eleventh-century stone statuary here depicted sexual situations, some of which seem startlingly bold even today.

In the West, we don't see sexual imagery in public locations—certainly not in places of worship. But the statues Burt saw, of men and women engaged in a variety of sexual positions, many quite acrobatic and imaginative, were on the outside walls as well as the insides of Hindu and Jain temples built a thousand years ago. Burt wrote in his diary, "I found seven Hindoo temples, most beautifully and exquisitely carved as to workmanship, but the sculptor had at times allowed his subject to grow a little warmer than there was absolute necessity for his doing."[1] There were once 85 temples at Khajuraho, and 20 of them survive to this day; they are both Hindu and Jain.

All temples in this area—"a stone's throw away from each other," as Burt wrote—have statues adorning their walls. These sculptures depict scenes from everyday life as well as images of deities. But the erotic scenes strongly dominate these temples because they are so explicit and so unexpected. These are almost life-size statues and friezes carved in gray, yellow, or reddish stone showing people in every imaginable sex act. At one temple, high above the observer, a man and a woman are supported by two seminude women; he is held upward, facing away from the viewer; below him and facing us is the naked woman, her head on the floor; her legs are spread above, and the two of them are attached in their genital region. At another temple, at eye level, there are statues of several elephants led by people, and on the right, unexpectedly, is a statue of a man mounting a woman from behind—oblivious of the elephants. At another temple, an entire panel depicts men and women engaged in several different positions of oral sex, as well as a combined oral and vaginal sex act by one woman with two men.

To date, no good explanation has been proposed for this unusual art. Tour guides lecture about the Kama Sutra to naive tourists who venture here, and magazine articles and tour books suggest that some of the sexually explicit statues represent the Hindu god Shiva and his Shakti, Parvati. The dreaded destroyer of worlds, if this is true, is only interested in his consort's body; and anyway, many of the temples are not Hindu but Jain. Some of the more scholarly sources conjecture that the imagery may have served as fertility symbols. But nobody really knows the answer.

As inexplicable as the erotic art of Khajuraho was, so was the discovery more than a century ago of a piece of complex mathematics in this location. I had gleaned hints of it in an old book on the history of mathematics. David Eugene Smith says, "[A magic square] appears in a Jaina inscription in the ancient town of Khajuraho, India, where various ruins bear records of the Chandel dynasty (870–1200)."[2] I had originally thought that Hayashi might have been led here by this reference; instead, it turned out that he had read the very first historical description of this curious mathematical piece, written much earlier. Hayashi had possessed the actual announcement of the discovery of this ancient magic square in the notes made by the prominent British archaeologist Sir Alexander Cunningham, who in the 1860s found the mathematical inscription by the entrance to a Jain temple.[3]

I spent several hours visiting all the temples of the compound, but nowhere found an inscription. Where was Hayashi's magic square? I asked every tour guide I met—nothing. Then a French tourist who had overheard me ask his guide said, "I think I may

Takao Hayashi in the former summer palace of the Maharaja of Mysore (with his son Makoto) in 1983 while researching the history of mathematics in India.

have seen some ancient numbers by the door of one of the temples in the Eastern Group." This was all the way across town, in a more deserted and rarely visited set of temples.

I left the fenced-off compound and walked for half an hour, passing stray cattle, which roam freely in all Indian towns, feeding on garbage, and finally found the ancient temples composing the Eastern Group. The edifices here were mostly Jain rather than Hindu. I went from one temple to the next, followed by stray dogs and children dressed in rags asking for money. Otherwise, the grounds were eerily deserted. A wind blew in from the fields, kicking up swirls of dust. I saw nothing. Then I came to the Parsvanatha temple, a Jain temple built in the mid-tenth century CE. The doorway was framed with erotic statues. A man, or perhaps a god, stared amorously into the eyes of the woman or goddess in his arms, her head turned up to him. His left hand fondled her ample breast.

And there on the right, inside the doorway, I finally found what I had come here for—the numerals Hayashi had seen 40 years earlier but could not remember exactly where. It was a magic square, with Hindu numerals (some of which are like our own and some so different they are not recognizable by a nonexpert), inscribed on the door of this thousand-year-old temple. This magic square was a four-row by four-column square with the following digits (here transcribed using our modern numerals):

7	12	1	14
2	13	8	11
16	3	10	5
9	6	15	4

Now notice some amazing facts: The sum of every horizontal row is 34; so is the sum of every vertical column, the sums of the two diagonals, the sums of all two-by-two squares at the four corners of the larger square, and also the sum of the central two-by-two square. The temple bearing this curious inscription is definitively dated by another inscription to 954 CE. So as early as the mid-tenth-century, the people who built and worshipped here

*The magic square at the entrance of the tenth-
century Parsvanatha temple at Khajuraho.*

understood how to construct such sophisticated magic squares. The
Khajuraho magic square is one of the oldest four-by-four squares
(earlier three-by-three squares are known in China and Persia).

With its magic square and numerals found on site, Khajuraho
provides the greatest example of Hindu numerals from as early as
the tenth century. The numbers found at Khajuraho and at other
early temples in India suggest that numbers here may have originated in connection with religious needs and practices. For example, ancient Indian documents called the Vedas—dating from as
early as the second millennium BCE—specify the sizes of temples
and the numbers of animals to be sacrificed; all are represented nu-
merically, and it is perhaps for this reason that the earliest Hindu
numerals appear in ancient temples.

Also, these numbers have been identified in one of the most
curious early documents ever discovered. Interestingly, in 1514, the

The decorated façade of one of the ancient temples of Khajuraho.

German artist Albrecht Dürer, who was fascinated by numbers, made a celebrated engraving called "Melancholia," which featured at its upper right corner a four-by-four magic square:

16	3	2	13
5	10	11	8
9	6	7	12
4	15	14	1

This, like the Khajuraho magic square constructed almost six centuries earlier, is a "normal" magic square, meaning that all numbers from 1 to 16 must appear on it, and its sums are all 34. But while the Khajuraho magic square is surrounded by smiling naked or seminude figures engaged in carnal pleasures, Dürer's magic square is placed next to a melancholic, solitary, and fully clothed female figure. This is one more example of the differences I perceived in logic—and outlook—between East and West.

The Khajuraho inscription not only shows that the tenth-century Indians were adept at this kind of magic-square arithmetic, but it also showcases the numerals they used at that time (shown in the picture of the Parsvanatha magic square) and the correspondence to our numerals. Which numbers are the same and which different? How did the Hindu numerals become our own? And how did they change?

6

THERE ARE DIFFERENT KINDS OF LOGIC—NOT JUST THE
Western, "linear" kind of logic. Though—very broadly speaking—
in the West religion often seems to be antithetical to sex, in the
East religion and sex are part of one grand celebration of life and
all its pleasures. Mathematics is linked to both sex and religion, as
the placement of the Khajuraho magic square and the erotic imagery in religious temples suggests. In fact, sex embodies perhaps the
greatest mystery of life, and mathematics—the abstraction of logi-
cally based processes we perceive in nature around us—is arguably
the greatest intellectual mystery.

I wondered whether the ancient Jains and Hindus who built
these temples pondered such deep mysteries. How could they not
have, given the evidence they left us? Why are our lives so deeply
ruled by sex? And why is the universe fundamentally ruled by
mathematics? What is the secret to desire? And why do numbers
behave in such curious ways, as evident in a magic square and in
the remarkable way arithmetic works? These may well have been
some of the questions that Eastern peoples asked themselves in antiquity, and their answers might have led them to belief, and hence

to the establishment of their religions and the invention of their gods and the construction of places of worship, where they placed symbols of their greatest mysteries: sex and mathematics. This, at least, was my conjecture. Was I on the right path?

The erotic statues of Khajuraho reminded me of the previous time I had seen ancient erotic art. It was in Pompeii, during another cruise I took aboard my father's ship when I was 14. Laci couldn't accompany me this time because my father had asked him to supervise the loading of specially roasted Italian coffee. The company insisted on serving its passengers only the highest-quality Italian coffee, so this commodity was always loaded onto the ship at an Italian port. My mother and sister went shopping, and my father stayed aboard. So the wife of the chief engineer, Ruth Chet, an attractive and sophisticated 32-year-old, accompanied me on the visit to Pompeii.

We arrived at the archaeological site and visited the antiquities, and then entered the special exhibits area, where erotic statues and frescoes found in the ruined city were on display. But the Italians had a strange, sexist rule in those days: Men of any age could visit the exhibit, but no women. I was naturally curious about this art and went in. My young age was no issue, but Mrs. Chet was barred, and despite her loud protestations, pleading, begging, and threatening, the guard would not let her pass.

In the exhibit hall I saw a statue of a small man with a giant erect penis in his hands, reaching almost to his neck, and couples on beds copulating in various positions. The women's breasts were often covered with strapless bras. These statues and frescoes were all pre-Christian, as Pompeii was destroyed in 79 CE, although the covered breasts might demonstrate a degree of modesty even in sexual situations. Once Christianity was adopted in the West, the use of erotic imagery declined drastically. This is in contrast with what was happening at the same time in India. As a 14-year-old, I was deeply curious about the subject and, naturally, also very embarrassed by it. And Ruth Chet, being barred from the display, took her frustrations out on me. As soon as we returned to the ship, she ran to my father. "Your son has a dirty mind!" she cried. "He went into the pornographic exhibit—and they wouldn't allow me to go in." My father laughed.

At 14, I was intensely shy looking at the Roman art. But now, at Khajuraho, I was a mature adult on a mission to discover ancient numerals. The mysterious suggestive statues of Khajuraho were in a sense like the mathematical objects that hid in their midst. Thinking about the similarities and the contrasts between the two assemblages of sensual art, I came to the conclusion that Eastern peoples of the tenth century had no hang-ups about sex and sexuality. The freedom exhibited by the Khajuraho statuary evidences such openness and sheer excitement about life and its pleasures that I felt certain it pointed to a fundamental difference between East and West. I wondered whether this disparity of views was somehow connected with the fact that Eastern logic is different from the usual Western way of thinking, and whether both relate somehow to the ability to abstract numerals out of the void and thus create a number system so powerful that it would one day take over the world. In the East, sex and logic and math seemed to be related.

We tend to think that our Western logic is the only valid kind of logic. A few years ago, I became frustrated that my sister,

Ilana, who had been diagnosed with breast cancer, was not making what I believed to be logical decisions. She shunned Western medicine in favor of Chinese qigong as her only treatment for the disease. In desperation, trying to understand how anyone could be so "illogical," I bought the book *Logic for Dummies* by Mark Zegarelli. He says that Aristotle was the true founder of classical logic. I read,

> For example, here's Aristotle's most famous syllogism:
>
> Premises:　　　All men are mortal
>
> 　　　　　　　Socrates is a man
>
> Conclusion:　　Socrates is mortal.[1]

So far, so good. But then he continues:

> The Square of Oppositions . . .
>
> **A:** *All cats are sleeping*
>
> **O:** Not all cats are sleeping
>
> 　　Aristotle noticed relationships among all of these types of statements. The most important of these relationships is the *contradictory* relationship between those statements that are diagonal from each other. With contradictory pairs, one statement is true and the other false.
>
> 　　Clearly, if every cat in the world is sleeping at the moment, then **A** is true and **O** is false; otherwise, the situation is reversed.[2]

But this disagrees with the Buddhist idea expressed by Nagarjuna: Anything is true, or false, or both true and false, or neither true nor false. His statement implies that there could be situations where the opposite of an assertion could be as true as the assertion itself. How is this possible?

To a Western mind, Nagarjuna's "true or not true or both or neither" may seem like absolute nonsense. True and not true are mutually exclusive and exhaustive states for any proposition. If something is true, then there is no way that it can be *not* true. In fact, this idea underlies what in mathematics we call the *law of the excluded middle*—which says that there is no middle ground between true and not true. In mathematics, the law of the excluded middle—the fact that true and not true are mutually exclusive and exhaustive states—forms the basis for much of the mainstream approach in proof theory. Proofs can be constructive, building step by step to a final, positive conclusion. But most often, we prove theorems by contradiction (because it is much easier and is often the only way we see how to do it). If we try to prove that something is true, we first assume that it is not true and then show that this assumption leads to a contradiction. That contradiction establishes the truth of the original proposition.

But the entire structure of proof by contradiction assumes that nothing in the universe can be both true and untrue, or neither true nor untrue. So if we disallow the law of the excluded middle, proof by contradiction would not hold, and many theorems in mathematics would be unproved and undecided. So, what's behind this puzzling statement attributed to the Buddha? And why should

Euclid's Stunning 300 BCE Proof of the Infinitude of Prime Numbers

Here is the best—and most ancient as well as most elegant—example of proof by contradiction. It goes back 2,300 years. It is the proof attributed to the Greek mathematician Euclid of Alexandria that there are infinitely many prime numbers. The proof proceeds as follows. Euclid says, "Let's assume that there are only finitely many prime numbers. Then there must be a largest prime number, after which there are no more primes and all larger number are composite (meaning they are products of prime numbers)." This makes perfect sense, right? If there are only finitely many primes, there must be a largest prime. Let's call this largest prime p. Now, Euclid says, consider the following number: $2 \times 3 \times 5 \times 7 \times 11 \times 13 \times \ldots \times p + 1$. This is the product of all the prime numbers, 2 through p, plus the number one. Is this new number prime?

If it is, then we have just exhibited a prime number greater than p. And if it isn't, then it must be divisible (by the definition of nonprime, or composite, numbers) by one of the primes 2 through p. Call that prime number by which $2 \times 3 \times 5 \times 7 \times 11 \times 13 \times \ldots \times p + 1$ is divisible q. But we see that this cannot possibly be true, since such a division will always leave the additional factor of 1 divided by that prime number, q, and $\frac{1}{q}$ could not possibly be an integer. So in either case, we have now exhibited a contradiction, which establishes the theorem.

we care? The reason I cared about this question was that I was convinced that all of this was tightly bound with the appearance of numbers—the mystery that had drawn me to the East.

While researching Buddha's logic—since I was so sure that it had something deep to do with the invention of zero and infinity—I came across an intriguing article by the American logician Fred

Linton of Wesleyan University. His curious paper actually explained the Buddhist idea of the four logical possibilities (they are called the *tetralemma* in Greek or the *catuskoti* in Sanskrit, meaning four corners) in the verse by Nagarjuna in a rational, mathematical way. Let's look at some everyday examples that Linton provides for situations where the additional two logical possibilities may hold: both true and untrue, and neither true nor untrue.

If you have a student, Linton writes, who is brilliant in mathematics but also has a knack for getting arrested in campus demonstrations, you might rightly say that he is both very bright and not very bright. A cup of coffee with just a small amount of sugar, Linton points out, could very well be described as neither sweet nor unsweet. Such examples abound.[3]

Apparently, Eastern thinking is more in tune with such gradations of truth and falsity, so the law of the excluded middle doesn't apply. In a sense, the strict interpretation that anything must be either true or not true may well represent a Western bias in thinking about nature and life. In an e-mail message to me, Linton provided more examples, of the opposite kind, which are obviously Western in their strict *either-or* bias: "You are either with me or against me"; "If you're not part of the solution, you're part of the problem"; and "Which will you have—tea or coffee?" Then Linton added, "I blame Aristotle for that!"

In fact, our Western logic does go back to Aristotle, who is famous for logical deductive statements such as the one above. But there are other kinds of logic as well, and they may apply in other situations and contexts. Eastern thinking modes seem to be more likely to accept differing ways of understanding the universe. But the question arises: Isn't it true that mathematics brings us only to the Western either-or kind of logic? Surprisingly, the answer is no.

Alexander Grothendieck (born in 1928) is one of the brightest and yet most troubled mathematicians of all time. Grothendieck had a penetrating vision in many areas of mathematics, including: the theory of measure—how we measure things even in the most complicated, abstract settings; topology—the theory of spaces and continuous mappings from one space to another; and algebraic geometry—the realm in which algebra and geometry merge, so that numerical information can be understood through geometrical forms. Grothendieck's entire oeuvre was motivated and driven by his quest to understand the meaning of numbers—including the deep concepts zero and infinity—as evident also in the formulation of the ten mysterious numerals that rule our world. His quest led him far afield, and he became the most celebrated mathematician of our time.

Then, at the peak of his career, during the 1968 student riots in Paris, he went a little crazy. The American war in Vietnam was at its height, and Grothendieck became so fervently antiwar that he traveled to Vietnam in protest. From then on, he was almost exclusively involved with political and environmental activism.

When asked to give a talk about mathematics, he would surprise his audience by refusing to speak about the intended topic, and instead turn his lectern into an antiwar, pro-environment pulpit. While most of his listeners were politically on his side, they also felt cheated: They had come to hear a mathematics lecture and not a political sermon, and they became disappointed with the man they had once admired.

Grothendieck then began to disappear for long periods of time and finally, sometime in the 1990s, made his final break from society. He is still living in hiding in the French Pyrenees, having withdrawn from the world. Reportedly, he is obsessed with good and evil and believes that the Devil rules the universe and has deliberately corrupted the speed of light from the nice round number of 300,000 kilometers per second to 299,792.458 kilometers per second.[4]

But long before he disappeared in his mountain hideaway, Grothendieck completely recast the field of algebraic geometry, as mentioned, and as part of that brilliant undertaking he invented a new concept: the *topos*. The topos is the ultimate generalization of the concept of space. Only Grothendieck could have the audacity, and the incredible facility with mathematics, to dare propose such a bold idea. Then, according to Pierre Cartier, a longtime member of the secret French mathematical association named Nicolas Bourbaki and a friend of Grothendieck (although he says that he has not seen him since his disappearance): "Grothendieck *claimed the right to transcribe mathematics into any topos whatever.*"[5]

This means that, pretty much, Grothendieck felt he had found a generalization that was so powerful that it allowed him to cast mathematics into any mold he pleased. He could view numbers not as mere numerals and abstract entities, but also as geometrical shapes; he could turn shapes into numerical quantities; equally, he could abstract both of them into entities that lived in highly esoteric mathematical realms that only expert mathematicians could visualize or understand—and then "do math" in these weird spaces that few have the ability to imagine. While a number is an abstract concept that can be symbolized by a set of signs (the numerals, arranged in various ways), Grothendieck took that abstraction to a whole new level.

It turns out that the mathematics of a topos, invented by Grothendieck, allows for a mathematically consistent and correct basis that justifies the Eastern kind of logic, as explained by Fred Linton.[6] Technically, our strict, either-or logic is necessitated by our reliance on the theory of sets as a basis for mathematics. This gives us the concept of set membership, which is unforgiving: An element is either a member of a set, or it is not; it cannot be both, or neither.

What Grothendieck (with the help of other mathematicians) did was to free mathematics from reliance on set theory and set membership. He employed something called category theory, in which there is no need for sets and membership laws. This freed him to define the topos, within which other logical systems, not requiring the either-or absoluteness, could justifiably exist. Thus it was that through Grothendieck's work, Linton was able to show that Nagarjuna's tetralemma had a perfectly valid mathematical basis. The topos places the Eastern logic with its four possibilities on the same solid foundation as our Western logic. And in the Eastern logic, a ruling concept is that of the void, or emptiness, or nothingness: the zero.

In Linton's topos, applied to the tetralemma, the opposite of "not true" is not the same as "true." This is the key to the Eastern way of thinking captured by Nagarjuna. In the West, not (not (true)) = true; this allows us to do proofs by contradiction and captures our strict way of thinking. But in Linton's topos, there is "true," there is also "not true," and there is yet a third thing altogether, called "not (not true)." To use Linton's favorite example, this logic applies perfectly in situations in which we say things like "The coffee wasn't unsweet"—meaning that it wasn't sweet but it was not unsweet either.

This reminds me of an occasion when, a few years ago, my publisher mentioned meeting the publicist they had hired to promote my book. He said, "This person wasn't unattractive." This is typical of Linton's in-between logic: My publisher didn't want to say the person was attractive, but he didn't want to say unattractive either.

Readers more familiar with fuzzy logic or quantum mechanics, where something can be in a mixture (probabilistic or otherwise) of two seemingly mutually exclusive states, may choose to view the Linton topos in this way. Linton's work proved that the Eastern logic of Nagarjuna and the tetralemma (catuskoti) has a solid mathematical foundation. Quantum computing, if it becomes a viable reality, may rely on related logical principles that are different from the usual logic we take for granted as the only one—as my friend C. K. Raju, then back at his university post in Malaysia, recently pointed out to me.[7] And I believe that the logic of the catuskoti is ultimately what led to the invention of the key numeral of our system, the zero.

7

IN INDIA, MATHEMATICS AND LOGIC—AND THE INTER-
mingling of mathematics and numbers with sex—are very ancient.
The earliest known texts in an Indian language are the four collections of religious hymns and rituals, mentioned earlier, known
as the Vedas. These were composed in an ancient form of Sanskrit called Vedic Sanskrit, also known as Old Indo-Aryan.[1] The
Rig-Veda is the oldest of these ancient documents and is believed
to have been composed as early as 1100 BCE.[2] This text already
displays a tendency toward extensive use of numbers, especially
powers of ten. Here are some of its verses:

No bad hymns am I offering by exerting my intellect
In praise of Bhavya ruling on the Indus
Who assigned to me a thousand sacrifices,
That incomparable king desirous of fame.
A hundred gold pieces from the fame-seeking king,
Together with a hundred horses as a present have I received,
I, Kakshivant, obtained also a hundred cows from my master
Who exalted thereby his fame immortal up to heaven.[3]

The historian of India John Keay notes that, in the ending verse of this hymn, "by substituting sexual terms for words like 'bliss' and 'creation,' it is just possible to grasp" a meaning that made an expert, B. K. Ghosh of Calcutta University, describe this hymn as obscene. We may view it as erotic:

> O resplendent lord, with brilliant radiance may you be
>   delighted.
> May your own bliss be consummated. Your delightful
>   creation,
> The holder of your bliss, is as exhilarating as the bliss itself.
> For you, the vigor, equally invigorating is the bliss,
> O mighty, giver of a thousand pleasures.[4]

We find in the *Rig-Veda* sexual imagery and also extensive use of numbers. According to historian of India John McLeish: "From the time of their earliest civilizations, the inhabitants of the Indian subcontinent had a highly sophisticated awareness of numbers."[5] McLeish further says that the people of Mohenjo Daro—one of the first known cities on the Indian subcontinent, part of the Indus Valley civilization, which flourished some 4,000 years ago—"used a simple decimal system and had methods of counting, weighing and measuring that were far more advanced than those of their contemporaries in Egypt, Babylonia, and Mycenean Greece. Vedic altars had to be built to exacting mathematical prescriptions; the correct dimensions and the right geometry were crucial."[6]

It appears that numbers in ancient India were invented for religious purposes very early in human history. While numbers were of a practical concern in the West—a necessity of banking, accounting, and everyday purposes—in the East numbers acquired a spiritual, religious meaning.

I read many sources on Indian mathematics. In the 1925 book on the history of mathematics by David Eugene Smith, I found the following:

> The early numerals of India are of various kinds. The earliest known forms are found in inscriptions of King Ashoka, the great patron of Buddhism, who reigned over most of India in the third century BC. The characters are not uniform and vary to meet linguistic conditions in different parts of India. Karosthi numerals are simply vertical marks; the Brahmi characters are more interesting. The Nana Ghat inscriptions, from the Nana Ghat cave, 75 miles from Puna, are a century after Ashoka's edicts.[7]

These last numerals from the Nana Ghat inscriptions include a 7 that looks just like our 7, and the 10 looking like the Greek letter *alpha*. They are shown below.

Numerals from the Nana Ghat cave inscriptions, showing 10 and 7 at center.

These appear to be among the earliest numerals that ultimately evolved into the ones we use today. Buddhist monks inscribed them on the wall of a cave high on a mountain in the Western Ghats. They lived in the cave and used it as a place of worship during the second century BCE. We also know that Buddhist travelers throughout Asia were the main conduit for the eventual spread of the base-10 number system across the continent. To visit the cave, one must make the arduous four-hour climb up the steep incline to the bluffs that hide the entrance to this underground Buddhist site. The Indian government has not done enough to preserve it, and the inscriptions bearing numerals that are the progenitors of our number system are now degraded through vandalism and neglect.

But where did the numerals go from there? How did they develop further, after their formulation during the time of Ashoka?

At the National Museum in New Delhi I found a large display explaining the evolution of the letters in Hindi and in other Asian languages. Not far from the display area I saw a working research center. I walked over and began a conversation with two researchers; I was surprised by what one of them said: "We don't like to admit it, but our written language really originates from Aramaic." This was unexpected. "Well," the middle-aged scholar wearing a jacket and bow tie continued, "India had long-standing trade relations with the Middle East and with Greece, and Aramaic—the lingua franca of the ancient Near East—influenced the development of our own script."

But I assumed that the numerals could not have come from there, since numbers used in the Near East were either the base-60 Babylonian ones or the Greco-Roman letter-kind of numerals. I made this observation to the two researchers, and they nodded in agreement and said that perhaps the numerals were indeed a genuine Indian invention, even if the progenitors of the written script had arrived here from the Near East.

In fact, the earliest numerals ever used were in all likelihood Phoenician letters, from which the Hebrew, Aramaic, and other Semitic alphabets evolved.[8] Phoenician is the oldest language in the Near East, and we know that Pythagoras traveled in this region and learned some of his early notions about mathematics from the Phoenicians and the Egyptians and their priests. Our letter A and the Hebrew letter *aleph* both derive from the Phoenician letter *aluf,* which means bull and was inspired by a stylized drawing of the head of a bull. This letter once stood for the number 1. As *aleph,* it is still employed in that role by some religious Jews today. As *alpha*, it was used for 1 by the ancient Greeks. The Romans then chose to use I for 1, II for 2, and so on, while the Greeks continued in their own alphabet with *beta* and *gamma* for 2 and 3, and so on, and the ancient Hebrews with *bet* and *gimmel*, and onward.

So the Nana Ghat inscriptions are extremely important because they provide strong evidence that our numerals are a genuine Indian invention made during the distant past, and then perhaps developed further while spreading around the world. These numbers were preceded by earlier Indian numerals, the Brahmi script seen in monuments of King Ashoka (slightly different from the Nana Ghat numbers); these, in turn, are related to the Kharosthi script, another alphabet employed in writing Sanskrit and regional

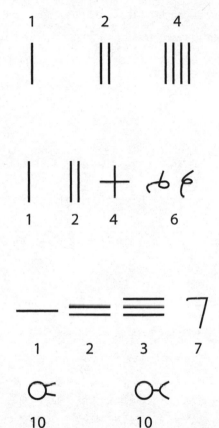

The Kharosthi (top, 3rd c. BCE), Brahmi (middle, King Ashoka's inscriptions, 3rd c. BCE), and Nana Ghat and Nasik Caves (bottom) numerals.

languages starting in the third century BCE in north India and Pakistan.

Noting that all these scripts bearing early numbers are Indian has made experts conclude that our numerals originate in India. But what about the zero?

8

ADMITTEDLY, C. K. RAJU IS A BIT OF AN EXTREMIST, AND
he probably enjoys the international notoriety he gets by making
provocative announcements at conferences around the world, such
as claiming that Euclid never existed or that Indians invented the
calculus long ago. But he is, in fact, fighting to right a wrong: to
bring back to the East some of the credit it deserves for inventions
made there, such as the invention of the numerals and the zero.
"Shunya—the zero—is clearly an Indian invention," he told me
with utter confidence in his voice when I saw him again. And indeed, the zero is an Eastern invention. But we can't say for sure that
it comes from India.

However, a Western bias has long influenced the history of
mathematics and science, and correcting historical misconceptions
is, and has been, a daunting task. Raju is one of the major players
in the enterprise to bring back to the East the credit it deserves in
this area.

One big problem with Indian antiquities in general—making
it so difficult to arrive at definitive conclusions—is the thorny issue
of dating. Determining the dates of artifacts and events in Indian

history has proved to be very hard: Some scholarly estimates of dates of items and documents have varied by as much as a thousand years. This problem has stood in the way of establishing an Indian chronology and especially has presented great difficulty in settling disputes about who invented what, and when.

A particular problem affecting the Indian history of numbers and mathematics has been a frequent lack of details in mathematical issues. Indian mathematicians have notoriously left out important parts of what a Western mind would expect in a mathematical proof—or in a reasoned quantitative argument—as well as dates of discoveries.

Even in the twentieth century, the mathematical prodigy Srinivasa Ramanujan, who produced many important results, left out key details of his work. Ramanujan was born in 1887 in the village of Erode, near Madras (now Chennai), in the state of Tamil Nadu in southern India. As a young man, he was able to derive hundreds of incredibly important and difficult mathematical results. When he sent them in letters to the prominent British mathematician G. H. Hardy, the latter realized that these amazing mathematical facts were written by a genius, but because Ramanujan provided no details or proofs, Hardy could not tell what was new and correct, what was not correct, and what was correct but had been known before. Hardy, nonetheless, was very impressed with the treasure trove of mathematical identities Ramanujan had sent him and said that "they defeated me completely; I had never seen anything in the least like them before." These equations had to be true, he concluded, because "if they weren't, nobody would have had the imagination to invent them!"[1]

Hardy was so taken with the unexpected mathematics of an unknown young mathematician from a faraway land that he invited Ramanujan to join him in Cambridge so they could work together. Eventually Ramanujan came to England, but he was gravely ill and, tragically, would not live long. A good example of his knowledge without proof is what happened between him and Hardy when he was in the hospital in Putney, in the United Kingdom.

While he lay there suffering from an unknown illness—now believed to have been a parasitic liver infection; he would die at 32, in the prime of his career—Hardy came to visit him. Not knowing what to say, he remarked, "I came here in a taxi with a rather dull number: 1729." At that moment, Ramanujan, weak as he was, jumped up in bed and exclaimed, "No, Hardy, no, Hardy! It is a very interesting number! It is the smallest number expressible as the sum of two cubes in two different ways." (Because $1729 = 1^3 + 12^3$, and $1729 = 10^3 + 9^3$.) Ramanujan just *knew* thousands of such facts about numbers and equations, and never bothered with derivations. He had no need to prove anything.

Indian proofs at times contained fewer details than might have been necessary for a complete understanding of a piece of work. Perhaps the convention on using a more terse form has something to do with a different kind of logic. In the case of the numerals, a similar phenomenon is evident. There are very ancient Indian documents that, if dated correctly, could probably demonstrate Indian primacy in the invention of numbers. But most of them—unless they are inscribed in stone—are undateable or undated. In the case of copper or bronze plates, of which many survive, the dates are usually placed off the main text, to the side, and could very well have been faked or added later. They are generally untrustworthy.

I went in search of one such plate, the Khandela inscription, which if found could shed important light on the story of the further development of the numerals in India. It might have been dated correctly and in a verifiable way—but nobody knew if it really existed. Still, I felt that I had to make an effort to find it.

RAJASTHAN. The name evokes images of desert horsemen in multicolored dress galloping on barren hills; fortresses by deep mountain lakes; mounted elephants marching in line to a fairy-tale palace; and of course snake charmers. From Khajuraho I flew to Jaipur, in eastern Rajasthan, in search of a copper plate I was not at all sure existed. While the Khajuraho airport is just a small airstrip and a hut, Jaipur's airport is larger and even has a restaurant or two. The city is on India's "golden triangle" tourist route, and the increased traffic has spurred recent growth.

Arriving late at night—later than expected due to dense fog in Delhi, where I had to change planes from Khajuraho—I sat inside the arrivals hall and looked out through the glass wall at people waiting to greet passengers. Then I caught sight of the limo driver holding a sign with my name; he had been sent by the hotel to pick me up.

In Rajasthan, for the price of an urban hotel room in any major city in the world, you can stay in a palace. And that's what I did—for my lodgings in Jaipur, I had booked the former coach house of the palace of the maharaja of Jaipur, which the present ruler rents out along with the entire palace with all its suites and rooms. The coach house was a bit cheaper than sleeping in the palace proper. Nevertheless, it was a charming place to stay, with Kashmiri carpets covering the floors, beautiful ebony cabinets, and a royal four-poster bed. It was quiet and peaceful, and I slept much better here than in the rundown Best Western in Khajuraho. The next morning, I hired the driver from the previous night to take me on a 50-mile drive to an old ruin to the northeast, where the Khandela inscription, a copper plate with early numerals and perhaps also a zero, was rumored to have been seen.[2] If it was still there, it was attached to the inside wall of a ruined temple.

We drove on a winding desert road and passed a lake with an island in its midst, and on it an ancient castle. On the shore, close to the road, a procession was taking place, with a succession of decorated elephants and camels carrying people. A small crowd gathered around an elderly man playing a wind instrument with his nostrils. We continued our slow ascent toward the ruins. During a stop on the way, I saw a handful of tourists standing around a snake charmer, the cobra moving its head as if to the rhythm of the piper's music.

We finally reached our destination, a ruined, deserted temple on top of a hill. The wind was blowing at the summit, kicking up dust. The ruined temple consisted of just two walls, and many stones that had fallen from the other walls covered the ground. But a walk around the remains of this temple revealed no copper plate anywhere. I spent two hours searching in likely locations, but there was no inscription anywhere. Many artifacts in Indian history have disappeared, and that gave me some comfort to soothe my disappointment. My driver took me back to the maharaja's coach house. Next I would go to find the earliest known zero in India. *This* inscription certainly existed.

Before leaving Jaipur, I went to visit the Jantar Mantar (observatory; literally "instrument formula") in this city—in which some prominent Indian mathematicians had worked many centuries ago. The Jantar Mantar of Jaipur is now used as a museum to explain early astronomy to the people of India and to visitors. I studied the sophisticated instruments on display. These devices predated telescopes, so no lenses were used, but they were remarkably advanced and could estimate with good accuracy the various angles to heavenly bodies; some were tracking instruments that could follow the movements of planets, the moon, and the sun over the entire year. I inspected the numerals shown on these instruments. What was on display here were the later devices used at this observatory, from the sixteenth and seventeenth centuries. Therefore, all the numerals were our modern ones, and included a zero.

THE EARLIEST VIEW that our numbers, with zero, came from India appears to have been proposed in a scholarly publication by the renowned German historian of science Moritz Cantor. In a publication of 1891, Cantor says, "This kind of conscious juggling with the notions of positional arithmetic with the zero is most easily explained in the home of these notions, which home for us is India and this we may affirm even if there is question of a second home. We mean if both notions were born in Babylon, of which there is great probability, and were carried over into India in a very undeveloped state."[3]

Louis C. Karpinski of the University of Michigan quoted Cantor's groundbreaking passage on the putative origin of numbers in his article in *Science* on June 21, 1912. He had translated Cantor's German into English.[4] Then his own article went on to discredit any notion that the numbers may have originated in Babylonia, in any form, because the Babylonians used a sexagesimal system, building their numbers using a very large base: 60. He pointed out that the Babylonians did not use any place-holding zero. His conclusion was that the numbers had to have originated in India. But what was the evidence that the numbers—and especially the all-important zero—were invented in India?

Neither Cantor nor Karpinski presented any such definitive proof. Karpinski said in his article that "[an] early document referring to the Hindu numerals has been published. This document is of prime importance because, being written in 662 A.D., it ante dates by more than two centuries the earliest known appearance in the ninth century of the numerals in Europe."[5] Surprisingly—for an article in the prestigious journal *Science*—Karpinski doesn't tell us what this document is. In fact, if such a document existed today—and if one could prove definitively that it was written in the seventh century—it would be one of the most important documents in the history of science. Equally surprising is his statement that the numerals arrived in Europe in the ninth century, again without any proof.

So in the meantime, lacking convincing evidence that the numbers including zero were of Indian origin, many scholars in Europe remained as skeptical as ever about any Eastern origins whatsoever—some claiming, as we will see, that the numbers and the zero were either invented by the Europeans themselves or by the Arabs.

Perhaps Karpinski was alluding to the famous Bakhshali manuscript. This mathematical document, written on birch bark, was discovered in the 1800s in the ground near the village of Bakhshali, not far from Peshawar in present-day Pakistan. It clearly is very ancient, and the bark on which it was written, 70 leaves of it, is so fragile that no one has been allowed to touch it for fear it will disintegrate. Today this ancient document is on display at the Bodleian Library at Oxford. Because the Bakhshali cannot be touched, no samples of it can be taken for radiocarbon analysis—which could reveal its actual age with excellent accuracy—and so we still cannot tell how old it is. Many scholars believe, based on linguistic and textual analysis, that it was created between the eighth and twelfth centuries CE, while others place it much earlier, anywhere from 200 BCE to 300 CE. But without radiocarbon analysis, it is not possible to date it definitively.

The manuscript contains a wealth of mathematical writings, from early equations to ways of estimating square roots to the uses of negative numbers. Most importantly, the Bakhshali uses a symbol for zero. If it could be dated to the second or third century, or even the fourth, it would establish that zero—and with it our entire number system—was invented very early in India. Should the British authorities ever allow a simple, hardly invasive procedure to be undertaken, in which a tiny amount of the bark is analyzed in a radiocarbon lab, we would know the real age of this extremely important artifact. Until then, the date of the most important Indian artifact in the history of mathematics remains very doubtful.

The British scholar G. R. Kaye was the first person to study the Bakhshali, at the start of the twentieth century, and he concluded that it was no older than the *twelfth* century. Therefore, he could argue for a European or Arab origin of our number system. He wrote, "The orientalists who exploited Indian history and literature about a century ago were not always perfect in their methods of investigation and consequently promulgated many errors . . . According to orthodox Hindu tradition, the 'Surya Siddhanta,' the most important Indian astronomical work, was composed over two million years ago!"[6]

Kaye clearly dismissed the research of Moritz Cantor, Louis Karpinski, and others who like them believed that the numbers and the zero originated in India. He continued in his article with a scathing attack on all who argued that our numbers come from India and mocked Indian date estimates, including those for the Bakhshali that placed it at an early era. His article continued, "In the sixteenth century CE, Hindu tradition ascribed the invention of the 'nine figures with the device for places to make them suffice for all numbers' to 'the beneficent creator of the universe'; and this was accepted as evidence of the very great antiquity of the system!"[7]

While I did not know it at the time, Kaye would play a major role in my story. In the meantime, working under the assumption that the zero—the key to our entire number system—was an Eastern invention, I asked myself why this was so, and inexorably I had to link it with the unique logic that I perceived in Asia. My thesis was that the number system we use today developed in the East because of religious, spiritual, philosophical, and mystical reasons—not for the practical concerns of trade and industry as

in the West. In particular, nothingness—the Buddhist concept of
Shunyata—and the Jain concept of extremely large numbers and
infinity played paramount roles.

THE EARLIEST ZERO IN INDIA is found in the city of Gwalior
southeast of Agra, famed home of the Taj Mahal. Gwalior's history is steeped in legend. In 8 CE, Suraj Sen, the ruler of Madhya
Pradesh, contracted a serious illness and was about to die. He was
cured by a hermit named Gwalipa, and in gratitude, Sen founded
a city and named it after the man who had saved his life. Gwalior
has many temples built over the centuries, and it has a famous fort
whose defense played a role in many conflicts throughout Indian
history. The fort was almost impenetrable; it stands on a high pla-
teau in the middle of the modern city, rising sharply to 300 feet
above its surroundings. This made it very hard for enemies to reach
it and breach its walls. In a Hindu place of worship called the Chatur-bhuja temple—"the temple to the four-armed god" (the four-
armed god in Hindu tradition is Vishnu, the powerful maintainer
of our world)—there is an inscription in Sanskrit on the wall that
records that it was built in the year 933 of a calendar whose starting point was 57 BCE. This makes the year it was built 876 CE.
The numerals 933 used here are surprisingly similar to our modern
numbers. The inscription also records that the land grant for the
temple had a length of 270 hastas (a measure of length). The 0 in
270 is the oldest zero that can be seen in India today.

So by 876 CE, the Indians had the crucially important use
of a place-holding zero at their disposal in a number system that
from our modern vantage point was perfect. Their system would

have enabled them to compute in a powerful, efficient, and unambiguous way. But would it be possible now to go still further back in time and find when the *first* zero had made its debut, the first exemplar of one of humanity's greatest intellectual inventions? I wanted to see it with my own eyes, to touch it, to feel it.

I left India with this remaining mystery unsolved. I learned much there, but nothing about where the key to the numbers—the primeval zero—came from, and when. If the oldest zero in India was from 876, then it was possible that it had come from Arabia—and had gone to Arabia from Europe—because the ninth century is well within the timeframe of extensive Arab sea trade. This was a time when Arab trade flourished, allowing for the possible transfer of goods—and ideas and information—across the realms the traders roamed, meaning between Europe and the East. Such transfer could well have taken place from east to west, or from west to east. And this was exactly the argument that the Western-biased Kaye had put forward in his lectures and articles. Lacking an earlier Eastern zero than that of Gwalior, Kaye's argument that our numbers with the zero numeral were either European or Arabic in origin could not be countered or disproved.

But if a zero could be found in the East that predated the emergence of Arab trade, this could provide strong support for the hypothesis that the zero was indeed an Eastern invention. This is why the Gwalior zero—important as it is—could not definitively determine who invented the most important component of our number system.

The oldest zero was of course the Mayan zero—but it was confined to Mesoamerica and went nowhere from there. And the

Gwalior zero was from the mid-ninth century—so it was no good as a historical landmark. If the Khandela inscription was ever to be rediscovered, it could bring the invention of the Indian zero down to 809 CE, as had been reported by people who claimed to have inspected it decades ago.[8] But because its date was still late, this wouldn't help much in finding a decisive, earlier zero that could settle the question of who invented the concept and the numeral.

9

WHEN I RETURNED FROM INDIA, IT SEEMED TO ME THAT
my research was at a dead end. The ancient Indians of the ninth
century had a zero, but this zero was concurrent with the Arab
empire centered at Baghdad—the caliphate—whose traders connected East and West. The zero could have been invented anywhere: in the East and brought west by Arab traders; in Europe
and transported to India through the same Arab naval commerce;
or invented by Arab mathematicians themselves and then taken
both east and west through Arab trade. If the Bakhshali manuscript were to be carbon dated and was found to be much more
ancient than the Gwalior zero, this might settle the problem. But
who was I to convince the stubborn British authorities to allow me
to carry out what they thought was an invasive analysis of a priceless artifact? Others have attempted to do so and failed.

I felt resigned to the fact that I might never resolve who invented the zero. I wanted to go on with my studies but had little
going for me at this point. *Should I look for another research project?*
I asked myself. Debra was very supportive, however, and suggested
that I keep trying. But I was losing faith in my ability to research

this topic any further. I simply could find nothing more about the zero. There wasn't anything I could do that would move my search forward; all my attempts were futile. I felt frustrated, angry, and depressed after devoting so much time to this search. So reluctantly I started looking for other research topics, encouraged by my friends and colleagues who felt I needed a subject other than the numerals to occupy my mind.

I found a compromise. The zero was beyond my reach, but I could look at other number systems and study them. The Etruscans—a mysterious Italic people obsessed with death and funerary arts whose culture flourished between the eighth and third centuries BCE in what is now Tuscany and a part of Umbria—had their own number system, which had never been fully deciphered. So I began to look at Etruscan numbers with renewed research vigor. Playing-dice made of bone had been discovered in Etruscan archaeological sites, and these provided hints about the shape of the numerals from one to six. All of these numbers were letters in the Etruscan alphabet, but the alphabet itself had not been deciphered, so we are not sure of the shape of all the Etruscan numerals—there has simply been a paucity of finds for us to be able to draw a clear-cut conclusion. I found this fact intriguing. And after a month of intense work, I made progress on discerning a similarity between Etruscan and Greek letters. For example, the Etruscans had no *g* sound, so they imported the Greek letter *gamma* to stand for their letter *C*. When the Etruscan civilization was subsumed into the Roman Republic in the first two centuries BCE, C came to stand for the number 100—thus completing a circuitous route from Greece to Etruria and finally to Rome. This was interesting research, but it wasn't the exciting quest for the primal zero.

Then something unexpected happened. While Debra and I shared a meal one day—she had come home for lunch to cheer me up—she suggested that I might look further into the story of the Gwalior zero, the oldest known zero in India, which I had recently seen. Unbeknownst to us at the time, this suggestion would bring about the needed breakthrough in my stalled project.

Following her suggestion, I looked again at the Gwalior zero, and to my surprise found an excellent online description of this artifact by the mathematician Bill Casselman of the University of British Columbia. So I called him up, out of the blue, to ask him to tell me more about Gwalior. He answered my call with alacrity, and through our pleasant, long conversation I learned that he had a surprisingly extensive familiarity with the history of numbers. It also turned out that he had been a doctoral student of the celebrated Japanese American number theorist Goro Shimura of Princeton, whom I had interviewed for my earlier book about Fermat's Last Theorem. This was a fortuitous connection to have discovered between us, and I hoped Casselman would become a friend.

He told me he was sure that an earlier zero than Gwalior's had been discovered in Cambodia and published many decades ago by the French archaeologist George Cœdès. Casselman didn't know more about this finding, he confessed, and suggested that I try to find out the whole story. I almost fell out of my chair when I heard him say this—I realized immediately that Cœdès had to have been the archaeologist that Laci had read about many years before. I had finally stumbled onto his trail almost by chance.

I was bewildered. How could I not have found it all on my own? Hadn't I been carefully looking into the history of the zero for so many months? And to embarrass myself further, I later even discovered that the book sitting right on top of my desk, *The Universal History of Numbers* by French researcher Georges Ifrah, had several references to the work of Cœdès—and I had completely missed them. I sat motionless for a minute, rubbing my eyes in disbelief. How could I have been so careless? And then I went to the fridge and poured myself a strong, icy drink. Once again, Laci had led me in the right direction—even if it had taken me four decades to find out where he was trying to point me.

I spent the following weeks working frantically to learn as much as I could about this little-known (to me, and to the general public—in scholarly circles he was well-known) French archaeologist and linguist who had changed our understanding of the history of mathematics—without, himself, being a mathematician. Cœdès was a fascinating character: a man with immense gifts of language and interpretation, who cared deeply about history and about righting the wrongs perpetrated by bigoted scholars. Cœdès discovered a much earlier zero than the Gwalior, analyzed and published it, and corrected our understanding of the history of numbers.

But it turned out that Cœdès's artifact with the early Cambodian zero had been lost. I now felt a need to find it again, to see it and bring it to the attention of the world: the first known zero—a testament to humanity's great intellectual discovery that led to the creation of our modern digitally ruled world, and proof that the East, not Europe or the Arabs, had it first.

I was now ready to get on the road again, and I finally had a starting point. But who was this scholar, George Cœdès, who made such a powerful discovery now presumed lost?

GEORGE CŒDÈS WAS BORN on August 10, 1886, in the elegant 16th arrondissement in Paris, just across the Seine from the Eiffel Tower, which would be erected when he was three years old. His father was a wealthy stockbroker. His grandfather, a Hungarian Jewish immigrant named J. Kados, was an artist who was determined to start a new life in France; he had left everything behind him on abandoning his native Hungary, including his name, which he changed to make it sound French. His grandson, George, would throughout his life maintain the ligature between the o and the e in Cœdès, as well as on the accent grave on the second e. And he insisted it be pronounced as *sehdehss*.

Raised in comfort in Paris, George eschewed a career in finance, a field his father had encouraged him to study, and decided instead to learn languages. His mother was born to a Jewish family in Strasbourg with deep roots in Alsace-Lorraine, a part of France bordering Germany where German is still spoken with some frequency today. Cœdès had good familiarity with German from his home and as a young man decided to study it. At 20, he spent a year traveling in Germany to master the language. He learned it so well that when he crossed the border back into France, the guards couldn't believe he was French and not German. When he returned to Paris, Cœdès enrolled in a language instruction program to qualify him as a language teacher.

He passed his national teacher's qualifying exams with ease, and in May 1908 received his license to teach German in French secondary schools. But life was not easy for a Jew of foreign origin in France of the early twentieth century. The country was still reeling from the infamous Dreyfus trial, which had polarized society and caused a resurgence of anti-Semitism among both the elite and bureaucratic circles. Many French schools refused to hire the brilliant young bilingual teacher. Being ambitious and single-minded, George refused to give up, and after trying for positions at many schools, he finally was appointed to teach German at the Lyceé Condorcet in Paris and set out on the career of a high school language teacher. But soon events would take him on a different path.

Shortly after he started teaching at the Lyceé, Cœdès was called to serve his country. In a sense, this was fortunate for him, since his call-up occurred in 1908, during a peaceful time not long before the outbreak of the Great War. But the French military was as anti-Semitic in that period as it had been for decades, making life difficult for the young officer. On leave, he could visit his doting parents and his school, where students remained attached to their now-absent favorite teacher.

One day in Paris, Cœdès decided to spend his afternoon in the Louvre. Anyone who has ever visited the Louvre has been overwhelmed by the richness of the paintings, statues, and artifacts on display in this museum—perhaps the world's greatest. On this early spring day in 1909, the 23-year-old George Cœdès entered the Louvre and went to the room displaying the Near Eastern Antiquities Collection. George abruptly stopped in front of the Babylonian stele depicting the Storm God. He studied the explanation of the display and was surprised to find that if he concentrated hard, he could deduce some connections between the words in French on the explanatory panel and the signs displayed on the stone artifact. He was even able to decipher the meaning of a few of the characters.

Stunned by what he had been able to do, Cœdès realized he had a rare gift. With some effort, he could understand the meaning of ancient languages whose letters and signs were carved in stone on millennia-old artifacts. By the time he reached the room housing the Southeast Asian Collection, he was hooked. He knew he wanted to spend his life decrypting such ancient writings.

Because of its colonial involvement in Southeast Asia, a region the French called Indochine, France had acquired a wealth of art and documents in its museum collections from Cambodia, Vietnam, Thailand, and Laos. Over the next few months until his discharge from military service, Cœdès spent every minute he could spare in some Parisian museum, armed with notebook and pen, copying writings in Old Khmer, the ancient Cambodian language that fascinated him the most. Six months later, Cœdès had become somewhat proficient in this language. As soon as he was released from the army, he enrolled at the École Pratique des Hautes Études in Paris to study Old Khmer as well as Sanskrit, the most important Indian language.

That year, he published his first scholarly article. It was a brilliant linguistic analysis of a Cambodian stele from the third century, in both Sanskrit and Old Khmer, and it appeared in the prestigious *Bulletin de l'École Française d'Extrême-Orient,* a publication edited in Hanoi in French colonial Vietnam and named after the educational and research institute the French had estab-lished throughout Indochina.

In the summer of 1911, Cœdès was awarded his doctorate from the École Pratique des Hautes Études and got his first job offer as a scholar. The École Française d'Extrême-Orient, which had published his first article, offered him a position as researcher in Hanoi. He immediately headed for Indochina.

Cœdès was a careful, ambitious, and determined scholar from an early age. When he studied a subject, he did it thoroughly and completely, often sitting for hours at his desk inspecting ancient documents—copies of inscriptions, pencil rubbings of stone arti-facts or steles—until he understood them completely. Slight, be-spectacled, and with a pallid complexion, he looked like a born bookworm.

Cœdès knew that the British scholar G. R. Kaye, whose pa-pers he had read, was a man on a nasty mission. He realized how deeply Kaye despised India, a country that had welcomed him as a researcher and even allowed him to be the first to study the Bakh-shali manuscript; Kaye had used his knowledge of Indian antiq-uities to argue that India followed the West in discoveries about mathematics. He had even used the discoveries of ancient Greek coins in India—proving Indian trade with Greece—to bolster his claims of European primacy, and his main issue for decades had been to oppose the idea that the nine numerals with a zero were invented in India.

Kaye held strong to the conclusion that since no datable ar-tifacts had ever been found with definitive dates for zero earlier than the ninth century, our numbers must have been imported into India from Greece, or perhaps other places in Europe, or Arabia. The fact that he was the first researcher to study the Bakhshali had endowed him with academic clout, and he used it aggressively to convince other scholars that he knew much about India and that Indians could not possibly have preceded the West in designing a number system. In the highly biased, anti-Eastern British scholarly community, Kaye found many allies, and his views prevailed. But Cœdès was determined to prove Kaye wrong.

10

CŒDÈS DEFINED THE CIVILIZATIONS THAT THRIVED IN
Southeast Asia more than a thousand years ago as "Indianized"—
he called them that because their peoples practiced Hinduism or
Buddhism, followed Indian social customs, held Indian cultural
values, and used Sanskrit in addition to their local languages. He
noted the strong influence of India on the kingdoms of Angkor,
Champa, and other dynasties of the region, which were in effect,
according to his view, cultural extensions of India.

By studying numerals found in ancient Cambodia and Indonesia, where his Indianized cultures thrived in the early centuries
of the common era, Cœdès was able to support his theory that
these civilizations used such numerals before their appearance in
the West in the late Middle Ages. Numerals were found on many
inscriptions from the eighth and ninth centuries analyzed by
Cœdès, as were others found in India. But research in India proper
also identified the key element of the numbers: the zero, discovered at Gwalior. Perhaps in the ruins of these Indianized civilizations, where so many stone inscriptions had been found, he might
discover a zero predating that of Gwalior, Cœdès thought. In the

meantime he spent much of his time studying the amazing culture that sprang up in the western part of Cambodia a millennium ago: the legendary civilization of Angkor.

Angkor Wat is the largest temple or religious building—taking into account *all* the cathedrals and basilicas ever built—in the entire world. This gigantic, beautiful, and architecturally unique Hindu temple was built in the eleventh century—about the time that Notre Dame was completed in Paris—near the town of Siem Reap (Siam victorious) in western Cambodia. Unusually, Angkor Wat faces west, the opposite direction from that of all other Eastern temples. This fact has puzzled scholars for decades; one explanation for it is that it was dedicated to Vishnu, the Hindu god of the west. Because it was considered a monumental event in the history of Southeast Asia, the establishment of Angkor—the city of the Khmer empire that built the great temple—in the ninth century helped historians define the entire chronology of Cambodia:

Pre-Angkor: from the first to the eighth centuries.

Angkor: from the ninth to the thirteenth centuries.

Post-Angkor: from the fourteenth through the twenty-
first centuries.

The region that gave rise to the powerful Angkorian civilization was divided in ancient times into various areas, each of them an independent or semi-independent kingdom. The kingdoms in the area of Southeast Asia during this time were Chenla, today's Myanmar (Burma); the territory that is roughly present-day Vietnam was called Champa; and Cambodia of the ninth century, referred to in Chinese records as Fu-Nan (and by others as "Water Chenla").

Like Khajuraho in India, Angkor Wat was said to have become lost to the jungle when the Khmer empire disappeared, and was supposedly rediscovered by nineteenth-century French explorers. At least that is the story one hears about Angkor in the West. In 1846, the French missionary Father Charles Emile Bouillevaux rediscovered the magnificent lost legendary city of Angkor, including its temple, Angkor Wat. Five years later, the French explorer Henri Mouhot visited the site and made detailed observations of Khmer remains in the area.

While it has no sexually explicit imagery, per se, Angkor Wat is adorned by thousands of alluring seminude female friezes of demigoddesses, or nymphs, called *Apsara*. Inscriptions here and elsewhere in Cambodia bear much numerical information in the form of dates and numbers of animals to be sacrificed, as well as measurements of lengths, widths, and heights. It seems that ancient southeastern Asian temples are also filled with symbols of sex and mathematics.

Based on his research on Angkor, George Cœdès wrote the definitive book on Angkor Wat, titled simply *Angkor*. Even today, his book is by far the most comprehensive treatise on the lost civilization of ancient Cambodia. Cœdès described Angkor Wat as representing the height of the Dravidian style in architecture, which originated in southern India millennia ago. The temple itself is said to represent Mount Meru, the mythical home of the Hindu gods and thus an earthly likeness of a divine abode.

Cœdès learned that linguists have determined that Old Khmer is thousands of years old, predating Thai. The indigenous cultures of Southeast Asia were advanced and had known how to make bronze for at least 1,500 years before contact with the civilization of India during the first centuries CE. Indian and Chinese influences in ideas and practices started here in the second and third centuries CE and then spread throughout the region. He learned that Chinese records from that time speak about Fu-Nan, in the lower reaches of the Mekong River, and Chenla, farther inland in the Mekong River basin. These references are in fact the only historical—rather than archaeological—ones we have for the civilization that grew in this region.

In the eighth century, Cambodia suffered from inner strife, but a century later King Jayavarman II, considered one of the greatest Khmer kings, united the kingdom and initiated the period we now call Angkorian. Over the next four centuries, the Angkorian civilization of Cambodia often occupied parts of Laos, Thailand, and the southern part of Vietnam.[1] It quickly gained power and influence to become the preeminent power in Southeast Asia.

As Cœdès's research focused increasingly on the numerical information he was finding in inscriptions from ancient Khmer temples, he developed a research purpose: to find an inscription with a zero sign that would predate Gwalior. If he could do this, he would be able to refute the theories of Kaye and prove, as he now firmly believed he could, that our modern number system originated in India or in an Indianized civilization such as that of Angkor or one before it. But he could find no such inscription. He did, however, locate and translate hundreds of the ancient steles of Cambodia. Then, in 1929, he came across the most astounding find in the history of numbers.

At a location some 300 kilometers northeast of Phnom Penh, in a wooded area by the Mekong River at a place where its flow is as mighty as anywhere, are a group of temples dating from the seventh century CE. These temples are now in ruins, but the kind of art found here—as evidenced in carved decorations of lintels and doorways, using small friezes and geometrical designs—is discussed as the unique "Sambor on Mekong" style of art. There is another Sambor, northwest of Phnom Penh, about two-thirds of the distance to Siem Reap, called Sambor Prei Kuk, which is larger, boasts its own architectural and artistic style, and also dates from the seventh century, and is thus defined as pre-Angkor.

A French archaeologist named Adhémard Leclère was working in the ruins of the temple of Trapang Prei at Sambor on Mekong in 1891 when he found two stone inscriptions written in Old Khmer. Much later these were brought to the attention of George Cœdès, who labeled them with the codes K-127 and K-128. (Many inscriptions studied, published, and catalogued by Cœdès were identified with a notation starting with K followed by a number.)

Cœdès started translating K-127. The inscription was almost intact. The top segment was partially broken, but the bulk of the inscription was perfectly legible and clear. What he read stunned him, as he realized that he had in front of him exactly what he had been looking for. This find was monumental—it included a zero! And the collection of temples that housed it had already been dated, using linguistic structures, to the seventh century CE—half a millennium earlier than the peak of the Angkorian empire and 200 years before the Gwalior zero. K-127 had come from inside the best-preserved temple, Trapang Prei. But Cœdès didn't need linguistic analysis to date K-127. Its date was written right on it. It read:

çaka parigraha 605 pankami roc . . .

Translated, it reads:

The çaka era has reached 605 on the fifth day of the waning moon . . .

Cœdès knew very well that çaka was a dynasty whose first king began to rule in 78 CE. So the inscription's date in our calendar was 605 + 78 = 683 CE. The writing and the numerals were in Old Khmer, and Cœdès was so good at this language that he had translated it within minutes. The zero—the first ever, as far as he knew—was clearly discernible and only slightly different in form from Indian zeros: Instead of a circle, it was a dot. Since 683 CE was two centuries older than the zero of Gwalior, Cœdès now had the proof that he'd been looking for.

Cœdès was very excited by this discovery, and eventually it would change our understanding of the history of numbers. In 1931, he published a paper still considered the most seminal ever on the origins of numbers. The article, "A propos de l'origine des chiffres arabes," appeared in the Bulletin of the School of Oriental Studies in 1931.[2] It reversed the world's understanding of the emergence of the so-called Hindu-Arabic numerals.

Cœdès wrote, "M. G. R. Kaye has insisted that 'We are forced to fix the ninth century A.D. as the earliest period in which the modern place-value system of notation may have been in use in India.'"[3] He then went on to completely destroy Kaye's hypothesis. He showed how the Sambor find changed this understanding: This first zero appeared two centuries earlier, and in an Indianized civilization in Cambodia, while not in India itself.

Kaye tried, when he was notified before publication of the Cœdès paper, to refute the significance of this find. He said it was only one example, and hence doubtful. But Cœdès was already prepared with additional proof to present in his article.

IN PALEMBANG, in the Indonesian province of southern Sumatra, another incredible discovery was made. On November 29, 1920, on a hill outside the town of Prasasti Kedukan, a roughly cut roundish stone was found containing the following inscription: "Congratulations. The year çaka 604 in the past, on the eleventh day, half-moon-lit . . . increase in the boat take of supplies to 20,000 . . . three hundred and twelve come to Mukha Upan . . ."[4]

There it was—a zero one year younger than the one from Cambodia (it is a year *younger*, rather than older, because the çaka era in Cambodia and in Indonesia differ by two years). This stone inscription, bearing the next-oldest zero we know, is still available for visitors to see at the museum in Palembang. It was made by another Indianized civilization that once thrived in Sumatra and on neighboring islands but then disappeared with few traces—even fewer than the remains we have of similar cultures on the mainland and in Cambodia.

The Sumatra finding gave Cœdès a second confirmation of the antiquity of the Indianized zero to definitively counter Kaye and his followers.

I was enthralled by the story of Cœdès's discoveries. But I knew that his first proof of the Eastern invention of the place-holding zero was now lost. Could I rediscover it?

11

I KNEW THAT IN ADDITION TO KILLING ALMOST 2 MIL-
lion Cambodians, the Khmer Rouge also destroyed or looted at
least 10,000 archaeological artifacts. What happened in Cambodia
in the late 1960s through the 1970s and beyond was akin to what
happened in Europe under the Nazis during World War II and
what happened in China during the Cultural Revolution. Starting
in 1968, Communist rule influenced by North Vietnam was estab-
lished over the country and was led by the Red Khmers, or, as they
became known using the French name, Khmer Rouge.

Headed by Pol Pot, the leaders of the Cambodian Communist government tried to impose a social engineering program
that would purge the country of intellectuals and bring about
a nationwide agrarian revolution—similar to Mao's dream for
China. Their methods, however, were far more brutal than those
of Communist China of the same years. Here the resemblance
to the Nazis comes in: Between 1974 and 1979—the apogee of
its rule—the Khmer Rouge tortured thousands of their fellow
Cambodians and killed a quarter of the country's population (it
has been estimated that between 1.7 and 2 million people died

in the infamous Killing Fields, out of a total population of 7.3 million). Even today, Cambodian society is still deeply wounded by the trauma perpetrated on the population more than three decades ago.

The Khmer Rouge was opposed to culture, art, science, and any intellectual pursuits, and it purposely destroyed much of Cambodia's cultural history, including archaeological artifacts, art treasures, and monuments. A visitor to Cambodia can see hundreds of ancient statues the Khmer Rouge wantonly defaced or broke to pieces.

I suspected that the Cambodian artifact K-127, bearing the first known zero of our number system, might no longer exist. It was likely destroyed by the Khmer Rouge, as were many similar finds of archaeological importance.

But I was determined to try to find this key stele even if the odds were against me. I wanted to refocus the world's attention on this immensely important icon of intellectual discovery in our distant past, which certainly changed the world; and I wanted to re-exhibit this unique archaeological find that had been used to counter the early twentieth-century Western biases in our understanding of history, lest these bigotries rear their ugly heads again. We need to understand history in order not to repeat it, and to remember it we need to see the proof of the first known zero. I was committed to finding this missing artifact.

I WORKED INCESSANTLY for several weeks, writing a research proposal I would send to the Alfred P. Sloan Foundation in New York. I laid out my argument about the importance of this find for the history of science and the need to try to recover it and study it further along with the civilization that had produced it. I explained that the idea of zero—both as a concept and as a place-holding numeral within our modern number system—is one of the most important in history. I described how the earliest known zero was discovered and later studied by Cœdès, and how he was able to use this monumental find from Cambodia to prove that the zero was an Eastern idea that later made its way to the West. I explained that the artifact had been lost and that rediscovering it should be of great value to society; it would recover for archaeology, scholars, and the general public the physical item that proves all of these facts about our history. The foundation concurred and awarded me a research grant to allow me to travel to Cambodia to pursue this study.

With the generous aid of the Sloan Foundation, I headed to Cambodia in early January 2013. This would be an arduous search that would take me into archaeology, mathematics, art, and also international intrigue and human chicanery.

TO PREPARE FOR MY TRIP, I felt that I also needed a deeper understanding of the concept of infinity as developed in the religions, philosophies, and mythologies of South and Southeast Asia. As seen in Khmer culture, the concept of an infinite ocean, on which the Hindu god Vishnu floats, is abstracted from an artificial lake. The very large artificial lakes, called Baray, created by a civilization that predates Angkor, represent in the local mythology a primordial, infinite ocean. Here, Vishnu reclines on Ananta's back, floating on this endless sea in cosmic sleep before Lakshmi wakes him up (see chapter 4). And then his offspring Brahma creates our world of space and time, springing out of a preexisting infinity.

We have, in fact, some historical evidence for this belief. According to Khmer interpretations of the writings of a Chinese visitor named Chou Ta-kuan (sometimes transcribed as Zhou Daguan), who came to Angkor in 1296, a large statue of Vishnu once loomed over the Baray at Angkor; water flowed from his navel, representing the birth of Brahma.

But in fact, Chou Ta-kuan's actual writings describe the statue as that of Buddha. Here is an excerpt from this unique report, the only historical description we have of the Angkorian civilization. It closely agrees with the archaeological findings at the site:

> The wall around the town measures almost seven miles. It has five identical gates, each one flanked by two side gates; there is one gate on each side, except on the east side, where there are two. Above each gate are five stone Buddha heads; the faces are turned to the west and the central one is decorated with gold. Elephants carved in stone are on both sides of the gates. Outside the wall there is a wide moat crossed by impressive bridges which lead to causeways. On both sides of the bridges are fifty-four stone demons, that, like the statues of generals, look mighty and terrible. The bridge parapets are of stone, carved in the shape of nine-headed serpents . . . A third of a mile to the north of the gold tower and even higher than it is a copper tower from where the view is truly impressive. At its foot are more than ten small houses. Another third of a mile to the north is the king's residence and attached to his sleeping quarters is still another

tower of gold . . . The eastern lake, about three miles beyond the east wall, is more than thirty miles in circumference. Rising above it is a stone tower and small stone houses. Inside the tower is a bronze statue of the sleeping Buddha from whose navel water constantly flows.[1]

Perhaps this originally described a statue of Vishnu, agreeing with the story that Brahma sprang from his navel. Or maybe Chou Ta-kuan was right and this was a Buddha statue. Hinduism and Buddhism come and go in the history of Southeast Asia: One religion dominates the other, and then the relationships reverse, over many centuries.

The Hindu destroyer of worlds, Shiva, is often depicted in art from this period (and also from the pre-Angkorian time) as riding on the back of a bull named Nandi. A very beautiful statue of Nandi was found at Sambor Prei Kuk from the seventh century, the time of Cœdès's inscription bearing the first zero. It is now displayed in the Cambodian National Museum in Phnom Penh.

Vishnu, on the other hand, rides on the back of Garuda, a mythical bird that transports him everywhere he wants to go. In Hinduism there is also a hint of heaven and hell: Yama is the ruler of the departed, and he decides who goes to heaven and who goes to hell after they die. This aspect of Hinduism is reminiscent of Christianity, where the existence of heaven and hell is so important, but in the East such notions are vaguer and less emphasized. There are other gods: Devi is the divine mother, worshipped by many; Ganesha is the half-elephant, half-human god believed to remove obstacles. He is revered by many Indian entrepreneurs.

Surya is the sun and Chandra the moon, viewed as gods, both reminiscent of animistic religions that may have predated Hinduism, Buddhism, and Jainism.

They also remind us of the religion of ancient Egypt with its sun god, Ra, and Mesoamerican religions as well, where the sun and the moon play central roles. In the entire region of Egypt down to Sudan—the realm of both Upper and Lower Egypt of antiquity—the sun god was the most important deity. In 2011, a hitherto unknown temple was discovered in Sudan's Island of Meroe on the Nile. The date of this temple was only very approximately estimated as 300 BCE to 350 CE, and its orientation was such that the sun's rays would penetrate it directly only twice a year. For this reason archaeologists have deduced that it was dedicated to Ra.[2]

We now come to Buddhism, which existed along with Hinduism in this region throughout history. Buddhism comes from India as well and has attracted many followers in Southeast Asia. The great king of Angkor, Jayavarman VII (ca. 1125–1218), another important and influential Khmer king, was himself a devout believer in Mahayana Buddhism and always expressed his faith in the compassionate Buddha who alleviated pain and cured illness. He placed Buddha at the central position among the gods, still maintaining the Hindu deities in subsidiary roles. Images of the Buddha from this time often include a *Naga,* the seven-headed cobra on which the Buddha sits. Another kind of Buddhism, Theravada Buddhism, is the one popular in much of Asia today. Buddhism is concerned with the easing of suffering, meditation, and the goal of achieving enlightenment, or nirvana. There are no gods here, but

Buddha serves as an example for a way of life to his followers. And of course, as I've mentioned, an important idea in Buddhism is the void, Shunyata, which may well have led to the mathematical idea of zero.

Jainism is a third religion that is still practiced in the East, in India and in other parts of Asia. This faith concerns itself with the transmigration of souls, and its adherents follow a very strict lifestyle. Because the soul of a deceased person may occupy the body of any presently living creature, Jains avoid eating all meats and make every effort not to kill even the smallest insect or other organisms. It turns out that the Jains, early in their history, became very interested in mathematics (as have Buddhists—the Buddha was himself a mathematician) and concerned themselves with extremely large numbers. They understood the concept of exponentiation and realized that exponents grow extremely fast (we even say today that something grows exponentially, with that same understanding). Thus, very high powers of 10, such as 10 to the power of 60, have been a recurrent concern for Jain thinkers from an early period in history—at least as early as the fourth century BCE, as attested in the *Bhagwati Sutra*.[3]

The three religions together give us concepts that did not arrive in the West until much later, in the late Middle Ages. These concepts are: zero, infinity, and finite but extremely large numbers.

I thus became increasingly convinced that the extreme number concepts—zero, infinity, and very large numbers—were intrinsically Eastern ideas. Their inception very likely required a kind of Eastern logic and an Eastern way of thinking. It was the East with its different views of the world that gave us the endpoints

of our present sophisticated number system. I found more evidence
to support this hypothesis.

Going back to scrutinize the writings of Nagarjuna more carefully, I found the following verses:

> Neither from itself nor from another,
>
> Nor from both,
>
> Nor without a cause,
>
> Does anything whatever, anywhere arise.[4]

This was a variant of the "true, not true, both, neither" logic, the
catuskoti (four corners), or tetralemma, which Linton had analyzed using the topos. But Nagarjuna then continues, in "Exami-
nation of Nirvana":

> If all this is empty,
>
> Then there is no arising or passing away.
>
> By the relinquishing or ceasing of what
>
> Does one wish nirvana to arise?
>
> If all this is nonempty,
>
> Then there is no arising or passing away.
>
> By the relinquishing or ceasing of what
>
> Does one wish nirvana to arise?[5]

So here he comes back to the emptiness, Shunyata. Nagarjuna
seems to be concerned with the void in all his thinking, and this
is intertwined with his concern with the logic of the catuskoti.
Nagarjuna has written extensively on Shunyata because he saw it

as the fundamental concept in all of Buddhism. And in doing so he linked the logic of the catuskoti with Shunyata. Was he intertwining the two concepts to lead us to the idea of an absolute zero?

The Buddhist writer Thich Nhat Hanh is even more explicit about the idea of the void:

> The first door of liberation is emptiness, shunyata
>
> Emptiness always means empty of something
>
> Emptiness is the Middle Way between existent and
>
> > nonexistent
>
> Reality goes beyond notions of being and nonbeing
>
> True emptiness is called "wondrous being," because it goes
>
> > beyond existence and nonexistence
>
> The concentration on Emptiness is a way of staying in touch
>
> > with life as it is, but it has to be practiced and not just
> >
> > talked about.[6]

As I concentrated on these notions, I came to believe that I could even read the quoted verses above as saying: existence = 1, nonexistence = −1, and emptiness = 0. Emptiness was the door from nonexistence to existence, in the same way that zero was the conduit from positive to negative numbers, one set being a perfect geometrical reflection of the other along the number line.

But I now had to *find* the lost Eastern zero—if indeed it still existed. I knew that in 1931, George Cœdès was able to destroy Kaye's argument in his seminal paper that employed this zero.[7] In fact, Cœdès presented in his paper two newly discovered zeros: One from Palembang, Indonesia, dated to 684 CE, and the one-year-older inscription from the Khmer temple at Sambor on the Mekong. In the paper, the Sambor find was identified as inscription K-127. This *K-* notation, instituted by Cœdès, would become my main lead in searching for the artifact.

The importance of K-127, and of the Palembang inscription, is due to the fact that both zeros *predate any Arab empire in the Middle East,* hence obviating the possibility of a likely knowledge transfer through Arab trade with Europe; since no European zero is earlier than these two finds, the argument about whether the West or the East invented the zero was settled. In addition, the locations of the two early zero finds were farther east than India, and this made it even more unlikely that the zero had come from Europe or Arabia. For if it somehow did, why were there no earlier zero finds in India but *two* zeros—both two centuries older than Gwalior's—found thousands of miles farther east? I had to find the lost K-127.

But the whereabouts of the actual artifact were unknown. Cœdès's paper included a pencil-rubbing bearing the Khmer numerals 605, where the zero is presented as a dot, but no known photograph of the find was known. In fact, since the resurgence of Khmer Rouge violence in 1990 resulted in the plunder of many more artifacts beyond the original 10,000 they had demolished in the 1970s, scholars in the field had implicitly assumed that K-127 was lost. Did I have any chance of finding it and bringing back to life this monumental piece of evidence for one of humanity's most ingenious inventions?

All I knew was that in the 1930s, the Trapang Prei tablet (K-127) was placed in the National Museum of Cambodia, and not much attention was given to it. In 1975, six years after the death of George Cœdès, Pol Pot and the Khmer Rouge took over Cambodia and turned the country's museums and repositories of archaeological artifacts into junkyards. Much was either completely lost or could not be accounted for. I was headed to Cambodia on a mission to find the lost inscription.

12

ON MY TRIP FROM BOSTON TO CAMBODIA, I STOPPED IN
Israel to see my sister. After six years with cancer and no Western
treatment, Ilana looked great and said she felt well. This was a
real blessing and I was relieved to hear it. It focused my attention
again, and from a different angle, on the differences between East
and West.

Western medicine believes that cancer must be fought against
aggressively, mainly using toxic chemicals and intense ionizing ra
diation—both of which kill normal cells as well and weaken the
body's immune system. Eastern logic, with its milder, more holistic
approach toward health and illness—not necessarily ruled by hard
science and statistics like the West's—prescribes meditation, herbal
medicines, and more naturalistic methods. In the case of my sister,
at least, it seemed the Eastern way was working. I thought back to
the time when I believed Ilana's approach was illogical and how
it induced me to buy a book on logic to try to understand how
she could think so wrongly. Now I concluded that she did have a
logic—just not the linear, Western kind. And apparently her logic
was winning.

For an outing one afternoon, Ilana and I went to downtown Haifa to see the port, where over many years our father's passenger ship docked whenever it returned home. The old customs house, which once stood by a large gate to the harbor, was now gone. "You remember what was here?" she asked. I said that I did and was surprised to see it gone. The new entrance to the port had been moved to a different, less prominent location, and the old customs house razed to the ground.

We continued down that main street leading to the old harbor gate, through which so many passengers had passed once they had undergone a thorough customs inspection in the old wooden building where it was always unbearably hot and stuffy. Turning a corner, my sister pointed to a large electronics store selling knock-off iPads and iPhones and assorted clones. "In those days, if you remember, they used to sell stereos and transistor radios—the electronics of that time," she reminisced. Yes, I remembered, I said. "And that's where he would sell all of his smuggled goods, you know," she continued. "Who?" I asked her. "Laci," she said, pronouncing it as only a person of Hungarian origin could.

"Laci was a smuggler?" I asked, incredulous.

"Oh, you didn't know? Why do you think he stuck with our father for so many years? It wasn't love or devotion, you know: It was all about money."

My sister had stayed close to all things nautical even after my father was no longer at sea. She later worked for his old shipping company, Zim Lines. "The captain's suitcases are never opened," she said, "out of respect for him—you probably remember that. Well, Laci, as the captain's steward, was the one to take these suitcases through customs. When our father was alone on the ship, he had a small bag—he had no need for anything larger because his uniforms were always on board, and most of his civilian clothes at home, and he bought very little. But when we were with him, there were several big suitcases, and Laci would usually hide something in our mother's largest suitcase, take all our luggage through customs, where it always passed through unopened and uninspected, and later remove whatever he had hidden in it in the alley just behind this store, and sell it inside. This went on for years."

"Did Mom ever know about it?" I asked, still shocked by this revelation.

"Well, remember those beautiful Italian Radiomarelli speakers we grew up with? Where do you think they came from?" I had no idea, I said. "Your trusted math tutor Laci miscalculated once: the taxi that always came to take us home up the mountain was there too early one trip, and the suitcases had to be loaded into it. Laci simply had no time left to remove his contraband."

"Amazing," I said. She went on: "Well, he never asked for his speakers back—he couldn't, of course—and I guess Mom felt it was a just payment for having illegally used her suitcase probably for many years. Dad never listened to those speakers—he didn't know what was happening right under his nose, but he understood that none of us had ever paid for them. Laci made a lot of money and then returned to Europe. I guess he is there now, somewhere— if he's still alive."

This was a hard piece of news for me to digest. I felt let down. The man I admired so much as a mathematician, someone who had taught me a lot and imparted to me a love of numbers—was in fact a smuggler? I found this revelation hard to believe. But I knew that my sister had to be right, as she always remembered details about our life on the ship with a stunning precision, perhaps reliving on an almost daily basis a childhood I had long left behind. I found it hard to sleep that night before my flight to the East to look for the origin of numbers—a quest I began under the influence of someone I considered a great man. *Was Laci really just a smuggler?* I asked myself again and again throughout that night.

Waking up the next morning, I resolved that the new information didn't change much for me in terms of my lifelong quest. Laci may well have had a dark side. He stayed with our family all those decades because he was able to exploit his position to illegal financial gain. But he was still a great mathematician and mentor to a young boy—and for that I was still grateful to him. I was continuing my quest regardless of what Laci may or may not have been. The search for zero was far bigger than him. And it was mine and mine alone now.

Within hours, I was at Ben-Gurion airport near Tel Aviv, ready to board my flight and continue my odyssey. I tried to put what Ilana had told me the previous day out of my mind and to concentrate on the task ahead.

13

AFTER A TEN-HOUR FLIGHT, I LANDED IN BANGKOK.
Twenty-seven years earlier, on our honeymoon, Debra and I came
to this city. At the time, the airport was small and the terminal
was just a tiny building. I was in for a big surprise: Bangkok now
has one of the most modern airports in the world, Suvarnabhumi
(pronounced in Thai as *Suvanapoum*) International Airport. Going through immigration was a high-tech experience with cameras
and modern technology.

I grabbed my suitcase and headed into town by the newly
finished fast train that connects the suburbs with the city center.
From there, the Sky Train, which hovers 50 feet above Bangkok's
crowded streets, delivered me to the steps of my hotel, the Shangri-
La, on the eastern bank of the Chao Phraya River—the city's main
watery artery, connecting its various parts north to south.

In the morning, I left the hotel by foot and walked north along
the bank of the river. Peddlers hawking street food made progress
difficult, and the smells of grilled meats, dried fish, frying onions,
garlic, and various spices were inescapable. Every Thai person on

these busy streets seemed eager to sell me something: if not food, then a fake Rolex or a pornographic DVD.

Within a couple of blocks, the street life became quieter, and the only sounds I heard were of an occasional tuk-tuk passing by, its driver slowing down to ask me, "Where do you want to go, sir?" But I shook my head, indicating I preferred to walk. I turned a corner into an alley and soon found myself in the midst of the former French colonial center of the city, which still includes the French embassy housed in one of the historic buildings from the 1920s, the tricolor proudly flying above the Chao Phraya.

A neighborhood of a couple of city blocks' radius around the embassy comprised the colonial section of the city, which has not changed architecturally from how it appeared in the early twentieth century, when George Cœdès walked these very streets to his office in this quarter.

He had close links with the Thai aristocracy, developed through his work on cultural affairs and through personal friendships, which enabled his important archaeological work in the region. It didn't hurt that the crown prince of Thailand was one of his closest friends, and that he had married a Cambodian princess. For a time, Cœdès directed the Thai National Library and was on the boards of a number of cultural and scholarly organizations. He had access to all the archaeological artifacts discovered anywhere in the wide geographical area of Indochina.

In his lifetime, he translated several thousand steles and inscriptions of all sorts from Old Khmer and Sanskrit. Cœdès was the undisputed world expert on Southeast Asia and its history and archaeology, and his word, often spoken quietly, was authoritative. This would help him change the history of numbers.

Behind the Mandarin Oriental, Bangkok's oldest and most celebrated hotel, located on the east bank of the Chao Phraya, I saw what I had come here to find: an intact French colonial building with gray wooden balconies and shutters of the kind one often sees in the French Caribbean today. This building once housed Cœdès's office, and I walked in. The internal staircase, with black wrought-iron railings, was surely original.

Where there were once offices, this old building now housed mostly small galleries and art dealerships. They displayed alabaster Buddha statues from Myanmar, seventeenth- and eighteenth-century white opaque likenesses of Siddhartha returning to earth from his heavenly voyage, hence the smile one uniformly sees on these particular sculptures that are sometimes robbed from Burmese temples; sandstone heads from Cambodia, purported to be authentic Angkorian statues of Vishnu or Shiva or the great king Jayavarman VII (these statues, if authentic, had to have been smuggled out of Cambodia, or else brought out before the passing of the law forbidding their exportation in the 1970s); and many wooden, bronze, or gilded Buddha statues from Thailand and Laos.

In what looked like one of the best art dealerships in this complex, the Galerie Mouhot, I met its owner, a Belgian man named Eric Dieu. He was dressed in red slacks and an orange open-collar shirt and sported a large gold watch, which I recognized as belonging to one of the most expensive Swiss brands, a fact he seemed delighted I had noticed. He was clearly very successful but proved to be extremely knowledgeable as well. Whenever I asked him something about an item on display, he would pull out some old art book and explain the history of it from the official information in these art guides. He sold mostly to dealers and museums, he said, and was clearly talking to me for the pleasure of sharing his passion for the region's antiquities, not because he thought he could sell me anything. I was curious about how a professional dealer might respond to my search.

I had a problem, I told him, which I came to Southeast Asia to solve. "I am looking for a very important pre-Angkorian inscription, on which the entire history of numbers hinges," I explained. "It was found in the late nineteenth century at the site of Sambor on the Mekong."

Mr. Dieu turned back to his well-stacked bookshelf behind rows of statues and small artifacts. He searched for a moment and then pulled out a large guide to the seventh-century Sambor art style. He leafed through quickly and found a number of inscriptions dating from that period, although not the object of my search. He then sat down and thought for a moment and, looking at the book again, found the name of the art curator of the museum in Siem Reap, Cambodia. He wrote the name down for me: Mr. Chamroeun Chhan. "I would start my search here," he said. "This man should know something about your lost inscription K-127."

This was a good place to start. As I left the building and returned to the Shangri-La Hotel, I made a mental note to search the Internet for the electronic or postal address of the curator of this museum in Cambodia; it included in its collections several inscriptions that were similar to, and of about the same age as, Cœdès's missing stele K-127. Equally, I thought that it would be useful to search for any information on Cœdès himself: perhaps old letters or notes, which might contain hints about the inscription. The next day, I visited the French embassy for this purpose. Its officials were very courteous but could offer no helpful information.

I searched everywhere in Bangkok—nothing. An Internet search for the curator of the Siem Reap museum did reveal an e-mail address, however, and I sent Mr. Chamroeun a note asking for his help. But I was frustrated that I could find nothing more related to George Cœdès.

14

BUT I KNEW THAT CŒDÈS DID NOT LIVE ONLY IN VI-
brant, exciting Bangkok, working in the charming building I had
just visited. He spent his time in the field looking for inscriptions
to translate and study: in Cambodia, Laos, and Vietnam, as well
as in the Thai countryside and in the Cambodian capital, Phnom
Penh, where he lived with his wife, the niece of the king of Cambodia. And for many years he did some of his best work while sta-
tioned in a city far to the north, Hanoi. Here he headed the École
Française d'Extrême-Orient (EFEO), the French cultural research
organization of Southeast Asia. So after a few more days in Bangkok, I took a night flight to Hanoi.

I arrived at Hanoi airport around midnight and waited in line
for more than an hour for a visa to enter this land. Vietnam is still
a Communist country. The unsmiling immigration officers, sitting under a large picture of the hammer and sickle, reminded me
of this fact. And the room's harsh lights made me think of KGB
interrogation rooms I'd seen in old movies.

Leaving the terminal, my passport finally stamped with a
Vietnamese visa, I took a taxi and began an hour-long drive on

a barely paved road in a darkness so complete that I had no idea what terrain I was passing. I normally travel easily and rarely have fear of unknown places, but the extreme darkness and eerie quiet made me uneasy. I had no idea where we were headed and had to trust that the driver, who spoke no English or French, was not taking me to an isolated field to rob or kill me. He drove without a word, and after an hour I saw the wide, tall walls of enclosures that must have housed many apartments. We navigated through many empty streets and alleys, barely lit here and there by a few sparse streetlights.

Finally we arrived at the entrance to a massive structure—the old French opera house in the center of the old city. It was now a French-run hotel. Between high, massive marble columns supporting the roof was the glass door to the lobby. A sleepy attendant opened the taxi's door and grabbed my bag. I paid the cab driver with relief and walked in.

This was a curious mix of a Western hotel with modern comforts and an Eastern kind of service. At breakfast—a buffet with Western-style cereal and eggs but also Vietnamese noodle soups with traditional fish cakes—the waiter came to my table and, hearing where I was from, muttered, "Osama bin Laden was good." I felt similar anti-American sentiments everywhere I went in Vietnam.

I walked through the old French colonial center of Hanoi, searching for Cœdès's former office at the EFEO. The building, in the 1860s style, was still there, but no French cultural organization. One of the people I asked gave me the name and address of a French expatriate still living in Vietnam who might possess some of the documents the organization failed to repatriate to France.

I took a taxi out of town, passing fields where water buffalo grazed and peasants wearing the typical conical straw hats plowed small plots of land. We finally reached a slow-flowing river, and a boat rowed by a strong young woman carried me downstream to a village where the man, Pierre Marcel, lived with his Vietnamese wife. Pierre was middle-aged, stocky, and had a pock-marked face. He was friendly, yet behind the outward French talkativeness there was something guarded. Who was he? What was he doing here in the countryside? I had an uneasy feeling that he might have been connected in some way with the French security services, perhaps keeping an eye on things in a Communist country decades after France had to abandon this erstwhile colony. Or perhaps the previous night's unsettling taxi ride had made my imagination overactive.

Pierre and his wife invited me to a lunch of the usual fish cake and noodle soup. Then he brought out an old cardboard box full of yellowing documents. After digging through them for a while, I found something of interest: a copy of a cover letter by Cœdès for an article he had sent to be published. It wasn't an important find, but still it was exciting to get to some original Cœdès material, if not to discover something directly related to K-127.

I mentioned my goal to Pierre and he said, "If it comes from Sambor sur Mekong . . . Well, there is an Anglo expat living in Phnom Penh who might help you with this. He knows this site very well, and maybe something about the whereabouts of the lost inscription from Sambor. His name is Andy Brouwer. Look him up online, you should find an e-mail address there." I stayed for coffee and then went to find my boat and its rower, who, after having lunch with people she knew in this village, had waited for me by the river.

The trip back was somewhat longer since we were now going upstream. We passed many birds perched on little bushes on the banks—kingfishers, mostly, and cranes, all searching for food—and lush jungle vegetation on both sides of the river. In the village by the road, I finally found a cab after looking for one for half an hour; it was getting dark and cold. The taxi driver was gruff and untalkative and spoke little English. Traffic was dense on the one-lane road back to Hanoi. Passing a car or an oxcart along the way was a difficult and lengthy operation, and we had to do it often. When a car passed us and almost pushed us off the road into a ravine, I began to wonder whether this chase was worth risking my life for. *But it is—I will find that missing inscription,* I told myself resolutely. The next day I was back in bustling Bangkok, ready for my great adventure to Cambodia.

I STAYED IN BANGKOK FOR A FEW DAYS, CATCHING UP
on e-mails and tracking down Andy Brouwer. Fortunately, he had
a very descriptive and visually appealing website, providing good
information on a number of otherwise inaccessible archaeological
sites in Cambodia that he had personally explored and clearly enjoyed writing about. He seemed, from these Internet blogs, to be
a gregarious man eager to share his knowledge of the country and
its treasures. He responded to my e-mail query with interest and
offered to meet me anytime, so I bought a ticket for a flight to
Phnom Penh.

The next day I left behind the shiny concrete, steel, and glass
skyscrapers of Bangkok and flew to hazy, low-lying, and crowded
Phnom Penh. I checked into the InterContinental Hotel at the
southwest end of the city and dialed the phone number for Andy
Brouwer. At his suggestion, we agreed to meet for dinner at the
intersection of Wat Langka and Street 278. It seemed weird to
me that streets would be named this way, but that is how it is in

Phnom Penh. The InterContinental is on one of the few streets with an actual name: Mao Tse Tung Boulevard.

I took a cab from the hotel, and the driver found the address. Wat Langka is an old Buddhist temple south of the waterfront area of the confluence of the Mekong and Tonle Sap Rivers, where the royal palace and other key monuments are located. Street 278 is just a small street with cafes and bistros catering to tourists, where one can find beer and meat pies. I had over an hour before my meeting, so I went inside to visit the Wat—which means *temple* in most Southeast Asian languages. It had high white walls and a red pagodalike top, similar to many other temples found in Thailand dating from the eighteenth and nineteenth centuries.

There were wide, open entrances, and inside, by a large Buddha statue, several monks wearing saffron-colored robes engaged in worship. Nobody paid attention to a lone Western visitor. Surrounding the main temple building were arrayed perhaps 50 small Buddha statues made of wood and of little distinctive artistic quality—nothing like the striking Buddha images at Wat Pho in Bangkok, home of the giant, golden reclining Buddha.

Leaving the temple complex, I made my way down Street 278. Western rock music blared from a bar where a handful of tourists were nursing tall tropical drinks. Down the street was a gift shop stocked with the kind of gaudy souvenirs that are ubiquitous in Asia: small metal and wooden Buddha statues, carved images of Angkor Wat, imitation stone heads of the Hindu gods, and carved phalluses on key chains. I looked around for a few minutes to pass the time and then continued down the street. There was a small nameless hotel, with a bar with high ceiling fans and wooden furniture. I walked in and had an Angkor beer and a handful of peanuts. A few French and German tourists were sitting at tables outside, enjoying the late afternoon sun.

After sunset, as the streets grew darker, I headed back to the intersection of Wat Langka and Street 278. I immediately saw a medium-height Western man with light hair standing there; he looked at me and asked, "Amir Aczel?" I said yes, and he shook my hand. "Hi, I'm Andy Brouwer," he said. "So, where would you like to go?" I said it was his home turf, and he suggested a restaurant he liked. We walked a block, and I barely escaped being hit by a swarm of motorcycles that rushed toward us as soon as the light on the boulevard had turned green. Andy seemed unperturbed, being used to the Asian traffic. At the next corner, we entered the restaurant, which served a wide selection of both Western and Cambodian food.

We sat down and ordered. "I'm from England, originally," Andy said, "from a town between Birmingham and Bristol." I told him that I knew the area and asked him what brought him here. "Well, I worked at a bank for 31 years—I started when I was 16— and always dreamed of breaking loose and coming to Cambodia. Only Cambodia. In the 1990s, I started visiting here, a month of paradise and 11 months of daily grind. And after many such years I decided it was time to be in paradise all the time." So he moved here and started working for a travel agency. But on the side, he indulges his dual passions: adventures in the wild and soccer.

I told him I was looking for an inscription, originally found at Sambor on Mekong, studied by Cœdès, and now of unknown whereabouts, its existence made more doubtful by the fact that the Khmer Rouge had looted and destroyed so many artifacts. "Well," Andy said, "I once found an inscription, myself."

I looked at him, surprised, and he continued. "I was exploring an area about 30 kilometers north of Angkor, where according to an old French topographic map from the late 1800s, which a friend of mine had procured at a curio shop, there were some unexplored ruins. Nobody knew about them. You see, the French made all kinds of maps and notations of things they found, but when they left Cambodia in the 1950s, all was either lost or taken back to Europe. So I navigated using this map, with my moto-driver. I had talked with the village chief and he had told me that he or his people had no knowledge of any temple ruins at the location I described to him, but that if I found any, to please tell him about it. And he sent his chief of police with me.

"Everything was densely overgrown with vegetation. This was now a virgin forest. People had not ventured here in maybe a hundred years. We literally had to hack our way in, meter by meter, using machetes. There were mosquitoes everywhere, and at one point we saw a cobra slither away. It was very rough going. But that's the kind of thing I love to do. So after a few hours of this, we arrived at exactly where my old map showed that there were ancient temples. And, lo and behold, we found the ruins of several temples: old bricks piled on the ground, some walls barely standing, stone doorways with carvings, and a hole where some ancient looters once looked for treasure. We sat down on the ground feeling victorious. And the guy facing me was striking the ground with a stick as we were talking, and I noticed that it sounded like he was hitting a stone surface. So I asked him to stop, and I leaned down and cleaned the dirt from under his feet where his stick was hitting. And as I did this I uncovered a large stone on which ancient writings were inscribed. Maybe it had an old zero on it, I don't know."

"Well, zeros are hard to find," I said. "But how old was it? K-127 is from the seventh century."

"This was later," he said. "These temples were closer to the time of Angkor—ninth or tenth century. So, anyway, we go back and I tell the village chief what we found and he makes a record of it to tell the authorities about it. Then a couple of years pass and one day I am watching TV and there is this *National Geographic* program about ancient history and this man makes a trek to that very same temple site and 'discovers' my inscription!" I said something about how unethical it was, but Andy said, "Well, at least the village chief didn't just forget about it but did tell someone where to look, and now they have another inscription." He went on to tell me about other adventures, including one where the ruins he discovered were so visually attractive and the surrounding jungle so lush that a Paramount Pictures locations executive he had made a connection with came to Cambodia within two weeks of the discovery for a preliminary study. A year later, they shot the film *Lara Croft: Tomb Raider* at that site.

"How would you go about looking for K-127?" I asked Andy when he finished his story.

"I have some connections," he said. "Let me make some phone calls, and I'll e-mail you. I'm pretty sure someone I know can help you."

Andy Brouwer on one of his jungle adventures in Cambodia.

"There is one more thing," I said. "I am interested in meeting knowledgeable Buddhist monks who might know about the concept of the void and be able to explain it to me. You see, I think that the zero probably came from the Buddhist Shunyata."

"That's an interesting idea," said Andy. "And I do know where the best place for you to look for an answer would be. It's called Luang Prabang, in Laos. It's a city of Wats and Buddhist monks. This is where I would go."

I thanked him and called my driver, who took me back to the hotel through the evening traffic jams of few cars, many tuk-tuks, and thousands of motorcycles. When I turned on my computer at the hotel, there was already a message from Andy connecting

me with one of his friends, a Cambodian man named Rotanak
Yang who knew antiquities well. A short time later, Yang e-mailed
me and promised to get in touch soon with any information, or
names of people who might know something about K-127. In the
meantime he told me that at one time, in the 1920s and 1930s, the
inscription I was looking for was indeed held at the national museum in Phnom Penh, as I had gathered from reading Cœdès, but
that it had long been removed from there—where, he didn't know.

So the next day, I went to visit the Cambodian National Museum, not far from the Royal Palace and the rivers. A plaque by
the entrance explained how the museum was founded in the 1920s
with the French governor of Cambodia presiding over the cere-
mony and the king in attendance. I found that somewhat sad and
degrading. I never did understand how a European power could
come here to Southeast Asia, presumably by ships that could carry
only a limited number of soldiers, and conquer lands with millions
of people living in them. I knew it happened because of the crazy
colonial ambition of Napoleon's nephew, who became known as
the Emperor Napoleon III.

As president of France in a democratically elected adminis-
tration, Louis-Napoléon Bonaparte staged a coup against his own
government to become emperor. The French military, at his com-
mand, took over the vast region they called Indochine in 1865.
Five years later, this same Napoleon was brought to his knees by
the Prussians attacking Paris. I always think of that dumb joke
about this event: "Monsieur, table for 100,000?"

But of course the French also gave us George Cœdès, who did
much for the region and brought us an understanding of its history

and culture. Some of the exhibits in this museum mentioned his name. And a large plaque by the museum's exit showed the names and photographs of all its past directors. One of them, from just a few years back, was Mr. Hab Touch, a name I would encounter again. He and his colleagues had done an incredibly good job restoring a once-great museum, which the Khmer Rouge had turned into a junkyard full of excrement from the thousands of bats and pigeons and other animals that had made the deserted structure their home during that dark period.

The Cambodian National Museum today is one of the finest museums in the world. The statues displayed here, from Angkor and Sambor Prei Kuk and many other sites in Cambodia, are among the most beautiful sculptures anywhere. There are images of Vishnu, with his four or eight arms; Shiva, usually with a third eye; and there is one exceptional statue of Brahma, his four faces serenely staring toward the east, west, north, and south. There is a beautiful statue of Nandin the bull; the female statues of the goddesses and consorts to the three Hindu gods are outstanding, as are the *Apsara,* the nymphs that one finds everywhere in Indochina. About half the displays are statues of Buddha from different periods and locations all over the land. He is often depicted sitting on his nine-headed cobra, the *Naga.* And Naga images are found at many places in the museum and elsewhere in Cambodia. The entrance to Angkor Wat, for example, is decorated with the nine-headed snake on both sides of the path leading to the great temple.

I left Phnom Penh with a little more information than I'd had before. I returned to Bangkok—my headquarters in the search for the lost K-127—to await more e-mails or phone calls that might provide me with further information. The search was becoming difficult since I had to rely on the goodwill of strangers who might know someone who knew someone else with information. It took an incredible amount of patience to wait to be contacted by people I did not know, who provided information at their own pace and out of sheer kindness and a willingness to help a researcher from another continent.

16

IN THE MEANTIME, WHILE WAITING FOR E-MAILS THAT
might lead me on a path to the lost inscription, I took Andy Brouwer's suggestion to look for answers about the Buddhist void in
Laos. I called Debra in Boston, and she managed to get off work
at MIT for a few days and flew for 25 hours, changing planes in
Tokyo, to meet me in Bangkok. We had not been together for a
few weeks—longer than ever before—and I was impatient to see
her again. We spent the night at the Shangri-La, and the next
morning went again to the airport to board a Bangkok Airlines
turboprop plane on a flight to Luang Prabang. While this was a
work trip for me, Debra was happy to accompany me, knowing
there would be time for us to spend together in an exotic and interesting place, and she was generous with her time trying to help
me with my quest.

Luang means *capital* and Prabang was the name of a Buddha image made in the first century and given as a present to the
Laotian king by the people of Sri Lanka. This statue can still be
admired in the national museum in Luang Prabang, even though
the Lao royalty is now gone and the country is Communist. Like

in Vietnam, visitors see the hammer and sickle frequently in Laos. Apparently, official corruption is prevalent.

Our flight from Bangkok lasted two hours. As we descended for landing, we could see the red-roofed pagodas of the temples, nestled in lush greenery and surrounded by tropical rainforests and palm trees; later on the ground I identified ficus and giant tualang and dipterocarp, which can reach 200 feet and are often covered with lichens.

Arriving in Luang Prabang airport in the late afternoon on April 4, 2013, Debra passed quickly through immigration and waited for me in the hall beyond the inspection area. I was next. I handed my US passport to the official behind the counter; he studied it for a long moment, then came out of his booth and said, "Follow me!" Noticing Debra waiting, he said, "Your wife can come too."

He ushered us away from other visa seekers and into a private room with curtains covering its windows. Another official was sitting there at a desk, reading a newspaper, never looking up, with a fixed smirk on his lips. The man who escorted us in drew all the curtains tightly and then said, "Sit down please." He half-heartedly offered us coffee but quickly turned to confront me. "We require passport validity for at least six months, and you have only five. This causes us a lot of trouble," he said, looking me in the eye. Then he said, "You have to pay us $200." I was startled, but I knew I had no choice but to pay. If I refused, they might detain me indefinitely, or send me back on the next plane and charge me for the ticket.

I shot a look at Debra and winked, indicating that I understood what these corrupt government agents were doing. Then I reached into my wallet and handed the man two $100 bills. "I will not have the same problem coming out of the country?" I asked.

"No, don't worry," he said as he led us back to his booth, stamped my passport with an entry visa, and sent us out to take a cab to our hotel, solicitously explaining which taxi to take so we wouldn't get ripped off. It was an unpleasant introduction to what would otherwise be a fruitful and enjoyable visit to a fascinating country.

The cab took us to the Kiridara Hotel, perched high on a hill overlooking Luang Prabang. Before entering the open-air lobby to check in, we wandered through the hotel's well-manicured gardens. Breathing deeply, we were aware of the strong, sweet smell of jasmine and other tropical flowers we couldn't identify; there was also a mild smell of smoke from the burning fields in the distance. It was a stimulating mixture of scents, and not unpleasant. We settled into our comfortable hotel room and drank cups of smoked Lao tea, the kind the villagers in the Laotian highlands drink. The next morning we ventured into town.

Luang Prabang is a jewel, an interesting mix of perfect architecture left over from its French colonial past. It's one of the few towns in Asia where virtually all the buildings have been left as they were a century ago. These are two-story houses with second-floor verandas and wooden shutters on windows painted white or light blue; the lower-level living areas have French doors that open to the streets, with large ceiling fans reminiscent of the movie *Casablanca*. French cafes that wouldn't be out of place in Paris serve fresh croissants and superb coffee. In the mix of buildings there are some upscale stores that sell local handicrafts, some restaurants, and travel agencies offering adventure tours into the surrounding jungles and boat cruises down the Mekong. Flimsy bamboo bridges cross the Nam Khan River, a small tributary of the Mekong, and the streets are a mixture of tuk-tuks, tourists, and many saffron-robed Buddhist monks. At night the town turns into a market, with local vendors and rural villagers selling gorgeous Laotian silk, handmade jewelry, and dead cobras in bottles. A local restaurant advertises that its specialties are buffalo, deer, and crocodile steaks.

We avoided these delicacies and shared a meal of curried rice with vegetables—the curry was better than Indian curry. And we talked about the project. "What would you do if K-127 doesn't exist anymore?" Debra asked cautiously when we finished our dessert of crème brûlée.

I thought for a moment, and then said, "I will find it . . . even if I have to spend the rest of my life here in Southeast Asia."

"So I guess we may have to take Miriam out of school and move here?" she said with a smile. "And how many more bribes do you think you might need to pay to do this?" We both laughed. But I knew she would stand behind me. Somehow—perhaps for no objective reason—we both had that confidence that I would succeed in finding K-127, even if it was now no more than a heap of broken pieces of stone. Debra knew how much this quest meant to me, and I was touched by her deep support. But neither of us knew how ugly it would get toward the end.

However, what we came to Laos for were knowledgeable monks serving Buddha in ancient temples, the Wats of Southeast Asia. We made our way to the oldest and architecturally most impressive temple: Wat Xieng Thong, a steeply pitched wooden pagoda with elaborate glass mosaics and gilded decorations built in 1565. We strolled through the wide grounds to the temple overlooking the Mekong on one side and the town on the other. The monks were walking the grounds or chatting in small groups. While Debra was taking pictures of the old temple, I approached one of the monks and explained what I was looking for.

He led me to the most learned scholar in this temple, who was sitting meditating by an impressive Buddha image inside the old Wat. I took off my shoes, came in, and sat on a wooden bench in a corner, waiting. When the monk finished his meditation, I introduced myself and asked him my question: "What is the meaning of the Buddhist void, the Shunyata?" He looked at me and thought for a moment, then replied, "Everything is not everything."

This was a considered answer from someone who knew a lot, and I understood that I shouldn't take it lightly. He was not searching for the right expression or confused about the use of English words. He meant exactly what he said: "Everything is not everything." I had to ponder this curious answer for a while. But I knew what he meant. Perhaps to an Eastern mind, "everything is not everything" might be intuitive and obvious in some sense. A Westerner has to think about it, and then it becomes clear and reveals its great depth and meaning.

To explain what "everything is not everything" means, I need to appeal to the work of the great English philosopher and mathematician Bertrand Russell. Russell proved in the early twentieth century that there is no such thing as a universal set, a set that contains *everything* inside it, leaving absolutely nothing outside. There is no container in which the entire universe or set of universes all exist with nothing left on the outside: There must always be *something* remaining outside of any kind of enclosure.

This mathematical idea has profound implications for the structure of the universe: The universe, whatever it is, cannot be all there is. Russell proved this surprising mathematical finding by an ingenious argument. He said, "Let's consider sets that contain themselves and sets that do not contain themselves." For example, the set of all dogs does not contain itself as a member, simply because it is not a dog.

But the set of all things that are not dogs does contain itself as a member. Why? Because it is not a dog, and hence it belongs with the collection of all things that are not dogs. Then Russell asked himself, What about the set of all sets that do not contain themselves? Does this set contain itself as an element? If it does, then by definition it cannot contain itself, and if it doesn't, then it does contain itself.

Russell used this paradox to expose some of the problems with the then-emerging theory of sets. We now know that the theory of sets does not agree well with the basic Eastern logic of Nagarjuna and the tetralemma, and we've seen how Linton, using Grothendieck's work—which was based on categories, rather than sets—was able to circumvent the problem. "Everything is not everything"— there is always something that lies *outside* of what you may think covers all creation. It could be a thought, or a kind of void, or a divine aspect. Nothing contains *everything* inside it. I found this idea profound. But he went on.

"Here," said the monk, motioning for me to come closer and offering me a tiny stool about 12 inches tall and made of an embroidered seat and four little wooden legs. I sat down beside him. "When we meditate," he said, "we count." He looked at me intently. "We close our eyes and are aware only of where we are at the moment, and of nothing else. We count breathing in, 1; and we count breathing out, 2; and we go on this way. When we stop counting, that is the void, the number zero, the emptiness." Here it was, I thought: the Shunyata and the number zero all in one.

I was beginning to understand what I had come here for. Here was the intellectual source of the number zero. It came from Buddhist meditation. Only this deep introspection could equate absolute nothingness with a number that had not existed until the emergence of this idea.

The monk continued. "We are born, we grow and develop, we become a quantity. Then we die, and this quantity becomes zero. This is the secret to meditation, and to existence." I sat there for a while on the tiny uncomfortable chair, contemplating what the wise monk had given me. Then I thanked him and left.

Crossing the square inside the temple grounds I ran into a crowd of European tourists speaking loudly in French, Italian, and German. In their midst was a tall Caucasian man dressed in a yellow robe, with a long white beard and hair bound in a ponytail. He was hard not to notice. I walked over to him and began a casual conversation about the temple we both had just exited. Eventually, I got around to the question that I wanted to ask this Western man who'd adopted the dress of the East: "What does the Buddhist void mean to you?"

"I'm not a Buddhist," he answered. "I am a Hindu. I am from Bezier, in France, but have lived for 41 years in Chennai—that's Madras."

"Yes, I know it's the old Madras," I said. "So what are you doing in a Buddhist temple?"

"Just visiting," he laughed. "I live here now, temporarily. I am Jean-Marc," he said, smiling.

Debra saw me speaking to him and waited. I asked Jean-Marc about the Hindu gods and their meaning. "I believe," he answered, "that God is not in heaven. You see, Shiva is in me and in you."

"Then we are all the destroyers of worlds?" I asked. At that moment the crowd of visitors surrounding us noticed this unusual man wearing a robe, which was somewhat different from those of the Buddhists: darker and a tad greenish. They flocked to him with questions. He didn't finish answering me. I walked over to Debra and told her about our conversation.

"Maybe you'll see him again," she said. "He seems to know interesting things." We walked back into town and had drinks at a French cafe overlooking the river.

The next day, I ran into Jean-Marc in a local shop as we were walking down the main street. He was there with his young Indian male companion; they were purchasing two colorful Buddha images. "Wrong religion, isn't it?" I asked him. He laughed. I went back to our interrupted conversation from the day before, about Shiva being in all of us. "Well," he said, "Shiva is indeed everywhere; he is in all of us, whoever we may be." I thought of the one man I knew who had justly—and tragically—applied this idea to himself: Robert Oppenheimer. At dawn on July 16, 1945, just as the first atomic bomb exploded a few miles away from him and other scientists in the New Mexico desert, Oppenheimer ruefully recalled a verse referring to Shiva in the Indian epic the Bhagavad Gita: "I am become death, the destroyer of worlds."

"So you are interested in Buddhism," I said, pointing to the two Buddha statuettes in his hand, one painted red and the other green.

"Yes, absolutely," he said. "There are interrelations among the religions of the East, as you can see, for example, at Angkor Wat. The temple started as a Hindu shrine to Vishnu—you know about the remains of a tenth-century statue of Vishnu they had just discovered at the top?—and now it is an active Buddhist temple. So, there you have it. And I'm sure you've seen images of Shiva's mount, the big bird Garuda, everywhere in these Buddhist countries of Southeast Asia."

I nodded, then said, "Well, I want to ask you about the catuskoti, the tetralemma . . . I've been reading Nagarjuna."

"You don't need Nagarjuna to understand the tetralemma," he said. "That weird-looking logic—from a Western point of view—is as old as Buddhism itself. Nagarjuna is just one of its later interpreters. You should study its earliest manifestations in Buddhism. As a mathematician, you will probably want a philosophical analysis . . . maybe it will answer both of your questions. Why don't you both come to my place?" he offered.

We thanked him for the offer and walked with him and his friend down the quiet street toward the Nam Khan, the Mekong tributary. We reached the riverbank and descended down to the sandy shore of the river, where a precarious-looking bamboo bridge stretched to the other bank above fast-churning waters. We walked carefully, holding both bamboo railings until we reached the other side. Then we climbed up a steep, dusty path to a hut perched at the top of the hill. Our host opened the door for us. We sat down in his small living room, and he offered us some tea. Then he turned to the bookshelf behind him.

He chose one volume, opened it, and read aloud to us: "In early Buddhist logic, it was standard to assume that for any state of affairs there were four possibilities: that it held, that it did not, both, or neither. This is the catuskoti (or tetralemma). Classical logicians have had a hard time making sense of this, but it makes perfectly good sense in the semantics of various paraconsistent logics, such as First Degree Entailment. Matters are more complicated for later Buddhist thinkers, such as Nagarjuna, who appear to suggest that none of these options, or more than one, may hold."[1]

As he read, he sat lotus-position across from us on the mat, playing with his long gray beard and stopping every few minutes to sip hibiscus tea. His eyes suddenly closed in meditation. After some time he opened them again and continued: "Within Buddhist thought, the structure of argumentation that seems most

resistant to our attempts at a formalization is undoubtedly the catuskoti or tetralemma."[2]

The description showed how the catuskoti appeared very early
in Buddhist thought, as early as the sixth century BCE when the
Buddha, the Indian prince Siddhartha Gothama, lived. Jean-Marc
read to us what happened when the Buddha was asked about one
of the most profound metaphysical issues:

> How is it Gothama? Does Gothama believe that the saint ex-
> ists after death and that this view alone is true and every
> other false?
>
> Nay, Vacca. I do not hold that the saint exists after death and
> that this view alone is true and every other false.
>
> How is it Gothama? Does Gothama believe that the saint does
> not exist after death and that this view alone is true and
> every other false?
>
> Nay, Vacca. I do not hold that the saint does not exist after
> death and that this view alone is true and every other false.
>
> How is it Gothama? Does Gothama believe that the saint
> both exists and does not exist after death and that this
> view alone is true and every other false?
>
> Nay, Vacca. I do not hold that the saint both exists and does
> not exist after death and that this view alone is true and
> every other false.
>
> How is it Gothama? Does Gothama believe that the saint
> neither exists nor does not exist after death and that this
> view alone is true and every other false?

Nay, Vacca. I do not hold that the saint neither exists nor does
not exist after death and that this view alone is true and
every other false.[3]

The only way to solve the conundrum, according to the article Jean-Marc was reading to us, was to conclude that the last two possibilities, both true and untrue and neither true nor untrue, "are *empty*."[4] Jean-Marc looked up in triumph, and I knew that my hunch had turned out to be right: The catuskoti, or tetralemma, collapses. Once we insist on *four* corners, these corners vanish, leaving us with the empty set: the void, Shunyata, or simply zero.

The connection I had been seeking for so long between the unusual Eastern logic of the catuskoti and the void, leading to zero, was now clear. The only mathematical solution to the logical paradox of the four possibilities of the catuskoti was the mathematical empty set: the great void, utter nothingness, the ultimate zero.

"So there you have it," Jean-Marc said. "The tetralemma leads directly to the Shunyata." We all looked at him, and he continued. "Buddhism emphasizes the void—something you in the West do not have. If you want, this may be what you are looking for—the source of the zero in the East is as old as the Buddha himself, 1,600 years old."

We sat there quietly for a while and then I said, "Thank you. Maybe now you can tell me about infinity?"

He laughed and said, "Ah, that's too big—maybe for another day?"

"May I come to see you tomorrow?" I asked.

"With pleasure," he replied, and shook our hands. Debra and I stood up, and his Indian friend took us down the hill to the bamboo bridge.

17

DEBRA PLANNED TO DEVOTE THE NEXT DAY TO TAKING
photographs. We agreed to meet in the late afternoon at the same
cafe we had enjoyed the previous day, with its view of the Mekong.
In the meantime, I went to see Jean-Marc in his small hilltop home
to learn more about his view of the Eastern infinity. I felt he would
likely know much about the concept, since Hinduism was comfortable with the infinite. He was in a jovial mood and offered me
a bowl of green curry vegetables and rice. We sat down at his table
and ate.

When we were finished, he said, "So you want to know about
the infinite in Eastern philosophy?" I said that I did because I believed that both zero and infinity—the extremes of our modern
number system—had to have come from the East. "The Buddha
himself was a mathematician, you know," Jean-Marc said. "In early
books about him, such as the *Lalita Vistara,* he is described as being excellent in 'numeracy' and able to use his ability with numbers
to try to win the attention of Princess Gopa. Numbers, including
very large numbers and their limit of infinity, appear in that text
already. Then, of course, we have in Hinduism many references

to infinity: infinite time, infinite space, and so on. It is far more widespread in Indian philosophy of that time than it is in the West. In the West, you only have some vague notions of God being infinite—whatever that means. But you should definitely look at Jainism, a religion that began early as well. The Jains, in particular, were interested in very large numbers." He walked over to his bookshelf and pulled out a volume and leafed through it. Then he said that infinite quantities are mentioned in a Jain text called the *Anuyoga-vara sutra* (Doors of Inquiry), written two millennia ago. The infinite quantities there are derived through an operation called "multiple multiplication," which might have meant exponentiation. If so, it would imply that the Jains who lived two thousand years ago understood something very deep about infinity.

"This is stunning," I said.

He smiled. "Yes, the ancient Indians understood infinity at least 1,800 years before mathematicians in the West did."

"So you know about Cantor's work?" I asked, surprised.

"Yes, of course. I studied philosophy for many years, including the philosophy of mathematics."

What was so surprising—and something I had not realized before—was that what he told me based on the Jain text provided some proof of a real mathematical understanding of infinity so early, and so long before a great genius in Germany, the tormented mathematician Georg Cantor, was able to explain the same concepts.

Cantor was a mathematician at the University of Halle in eastern Germany in the late 1800s, where he single-handedly developed the mathematics of infinity. He had been a student at the University of Berlin, one of the most important universities in Europe at that time, studying under a mathematics giant, Karl Weierstrass, who contributed hugely to our understanding of the real numbers: the numbers on the real number line, which include both rational numbers (integers or quotients of integers) and *irrational* numbers (numbers, such as pi, that cannot be expressed as quotients of integers). Weierstrass, together with another German mathematician named Richard Dedekind, understood that irrational numbers had infinite, nonrepeating decimal expansions: things like 0.1428452396 . . . , as compared with 0.48484848 . . . Decimals that repeat, such as the latter example, can be proven to always equal a rational number, meaning that they can always be written as ratios of two integers—In this case, the number 16/33.

Nonrepeating decimals—the best example is pi = 3.1415926 5359 . . . —are never rational, meaning they cannot ever be written as a ratio of two whole numbers. Cantor extended this entire study to a profound and new understanding of actual infinity. He understood that the decimal expansion of an irrational number is nonrepeating and *infinite.*

He also understood something counterintuitive: There are various *levels* of infinity, meaning that not all infinite quantities are equally big. Though infinite, some numbers can be larger, in some sense, than other infinite numbers. And Cantor was able to prove mathematically that, while both sets are certainly infinite, the set of irrational numbers is of a *higher order of infinity* than the set of all the integers. That is, there are more irrational numbers than there are integers.

Cantor's concept of infinity was controversial within the mathematical community of his time, and his frustration at the reception

Cantor's Deep Mathematical Analysis

In one of the most brilliant proofs in the history of mathematics, Cantor was able to show that the rational numbers—fractions made of integers as numerator and denominator—are of the same infinite "size" as the integers.

Cantor understood that the operation of exponentiation was the lowest (and the only one we know) arithmetical move that could take an infinite quantity from one level to a higher level of infinity. Exponentiation is essentially a move to the *power set*—the set of all subsets of a given set. This is one of the reasons why Bertrand Russell's paradox is indeed a paradox: We cannot find a universal set because no set can contain its own power set! Let's look at an example, the set containing only two distinct elements. Let's call this set X and its elements *a* and *b*. Now the power set of the set X with only two elements, *a* and *b*, is the set that contains all subsets of X. It is, therefore, the set that contains: 0 (the empty set), *a*, *b*, and (*a, b*). These are all the subsets of X. We see that the power set is always larger than the set itself (because the set X itself contains only *a* and *b*). The power set has more elements, and the reason for this is that the power set has 2^n elements, where *n* is the number of elements of the original set. So for any set, the power set associated with it is always larger than the set itself. If there *were* a set containing everything, its power set would still be larger, obviating the assumption that the original set contained everything. "Everything is not everything," as the monk had told me.[1]

to his work, as well as the difficulties he faced in developing his theories of the infinite, contributed to years of mental instability. Cantor suffered bouts of depression throughout his life, underwent lengthy hospitalizations, and died in a mental institution in Halle in 1918. He explained infinity to the world. Cantor showed that the continuum of numbers between any two numbers on the number line must have 2^n elements if there are n integers (n is infinite here). We know what "grows exponentially" means—so you can see that raising a number, here 2, to a power that is *infinite* creates an exponentially larger infinity. Cantor thus showed that the order of infinity of the real numbers (meaning rational numbers and irrational numbers, like pi or e, the base of the natural logarithms) is higher than the order of infinity of the rational numbers alone, or the integers.

In any case, it is possible, based on what Jean-Marc had said to me, that the Jains of ancient India understood that exponentiation raises the level of infinity when an infinite number is exponentiated, and creates a really large number when the number being exponentiated is large but still finite.

"You see that the ancient Indians understood infinity almost as well as Cantor did in the late 1800s," Jean-Marc said.

"So, let me see," I said. "Zero comes from the Shunyata through the logic of the catuskoti; and infinity comes from Hindu, Jain, and perhaps also Buddhist mathematical and philosophical considerations that go back 2,000 years."

"Sounds reasonable to me," Jean-Marc answered, seeming to be distracted by something. He rubbed his forehead and brushed back his long, curly gray hair. Then, as if an afterthought: "But, tell me, then. Do numbers really exist?" He looked at me triumphantly, like a chess player delivering checkmate.

"That's the biggest problem in all of the philosophy of mathematics," I said.

"Yes, indeed," he answered.

"Numbers are our greatest invention, and zero is the capstone of the whole system," I said. "But whether they exist outside our minds, outside their role as a construct that helps us understand the world around us, is a big open problem. I've interviewed many mathematicians about their views."

"And what do they say?" he asked.

"The majority are Platonists: They believe that there is a Platonic milieu in which numbers exist, quite independently of people or animals or any physical reality. But others are divided on this question. What do you think?"

"As a Hindu," he said, "I certainly believe in an immanent, divine reality. As I've told you, I believe that Shiva is in me and in you and in everything and everybody. If we aren't here, Shiva still is—and so are numbers and mathematical and all spiritual essentials. There is a reality that goes beyond people, and it includes numbers."

I was impressed with his erudition, his Eastern kind of Platonism. It was definitely time for me to continue my search for the first-known tangible evidence in Asia of the discovery of the idea of a zero—either invented or deduced from a latent reality.

I thanked Jean-Marc for the meal and the fascinating discussion and went down the hill to the bamboo bridge. I paid the bridge-keeper a dollar toll to cross it back to Luang Prabang, and walked through town to meet Debra at our cafe.

We had dessert together. Outside Paris, Luang Prabang is perhaps the best place in the world for delicious French pastries. We shared a *tarte aux pommes,* and I had a smoked tea while Debra drank a cappuccino. We watched the sun set over the Mekong—glorious shades of red and orange through a filmy white mist over the river. I told her about my conversation with Jean-Marc. "He sounds a bit like Roger Penrose," she said, referring to a book we had both read, Penrose's *The Road to Reality,* which discusses the issue of whether numbers were invented or discovered. Then we looked at the pictures she had taken during the day and the ones she had just taken of the spectacular sunset.

We walked together to our secluded hotel on the hill overlooking the town. I had found what I had come to this enchanted town to discover: the source of the zero and the source of infinity, embedded in the millennia-old wisdoms of Buddhists and Hindus and Jains. I now badly needed specific information about K-127 and was eager to fly to Cambodia to look for the Khmer zero that George Cœdès had studied 80 years ago. I hoped the inscription had survived the vicissitudes of time and the ravages of the Khmer Rouge in Laos's neighbor to the southeast.

The next day, we packed our bags and took a cab to the small airport, still nearly empty despite the growing demand from tourists. People were talking about a new airport to be built, and the construction of a high-speed rail line from China in the north. Once these two projects were completed, the town of Luang Prabang would be full of Chinese and other tourists. We knew prices would go up, high-rise hotels would be built, and the peaceful atmosphere would likely change.

I was a little worried that I might be asked to pay a penalty again, this time in order to leave the country on an "improper" passport, but to my relief my passport was stamped and we were allowed to board our flight. We returned to Bangkok and Debra flew home. Our mini second honeymoon in the oriental gem of Luang Prabang was over too quickly. I remained in Bangkok and awaited information on the whereabouts of K-127.

18

JUST WHEN I NEEDED SOMEONE TO CLARIFY SOME OF THE
ideas I'd been pursuing, I got a call from my friend Jacob Meskin,
a Princeton-educated philosophy professor and expert on the religions of eastern and Southeast Asia. We hadn't talked for many
months and caught up on conversation about mutual friends and
about philosophy. I tentatively explained to Jacob my view about
zero, Shunyata, and the catuskoti. "That's an interesting connection," he said. "In fact, Nagarjuna does talk about the void as a key
principle, and of course the emptiness, the empty set if you will, is
a solution to the 'four corners'—we don't see too many other good
solutions here." He chuckled. "Buddhism has a lot of numbers in
it: the three marks of existence, the four truths, the eightfold path,
the twelve-link chain of dependent co-origination, and so on. But
tell me about the numbers. Why is the zero so important? I don't
really understand that."

I explained to Jacob how the use of the place-holding zero is
what allows the numerals to cycle; it is what enables us to use the
same nine signs (plus the zero itself) again and again—for different purposes. For example, we can use the numeral 1 to mean the

number one; but when we place a zero to its right, that same sign, 1, now means ten. A 4 alone means four, but when followed by two zeros it becomes four hundred. It means 4 hundreds, 0 tens, and 0 units. The existence of a place-holding zero is what gives meaning even to numbers that contain no zeros in them. The number 143, for example, could not be written this way if the number 140 didn't exist, and this number needs the zero as a place-holder for the empty units. Without the zero, none of these numbers and manipulations of such numbers would be possible. "That's interesting," he said. "Nagarjuna actually talks about the void being movable from place to place—just like your place-holding zero. Perhaps he understood that, too. I like to think about this as the little plastic toy that children play with: a square with numbers that can be moved around, but the numbers can only be moved because there is an empty space for one missing little number-square. This missing piece allows us to move the numbers around one at a time until they are in numerical order. So, you see, the void is everywhere and it moves around; it can stand for one truth when you write a number a certain way—no tens, for example—and another kind of truth in another case, say when you have no thousands in a number!" Seeing Nagarjuna's apparent view of the void, and perhaps the zero as well, as a dynamic, movable piece was certainly intriguing.

I mentioned my feeling that in India, math and sex and religion were intertwined. Jacob responded: "Forgive me, I am sure you've thought of this, but there is, it seems to me, a line of connection in this conversation here between numbers and . . . ahem . . . sex. It's a bit weird, but here it is. Nagarjuna expresses it toward the end of chapter 24 of his *Mula-madhyamaka-karikas* (Fundamental middleway verses). In that chapter he imagines a critic attacking the view that he, Nagarjuna, has been presenting in the preceding 23 chapters. Nagarjuna imagines a critic accusing him of something like nihilism. This imaginary critic says, in effect, 'Hey Nagarjuna, you've made Buddhism into the teaching of Shunyata. But that means that everything is empty. And isn't that just like saying that nothing is really true? And doesn't that mean that everything the Buddha said isn't really true?'

"Nagarjuna's answer is fascinating. He says that his critic has everything backwards and that it's only *because* everything is in fact empty (Shunyata) that everything actually works, including the truths the Buddha came to state. Sort of like 'without Shunyata, nothing works; with Shunyata, everything works.'" Then Jacob paused. "I will send you an Internet translation of some chapters from Nagarjuna's *MMK,* and it includes chapter 24. I'd love to go over it with you if you'd like. But the bottom line is that if everything really did possess an eternal, unchanging character—an essence, *sva-bhava* in Sanskrit—then the basic claim of the Buddha, namely that everything arises only via a complex set of cooperating and conjoined factors, could not be true. The Buddha is insisting that everything is endlessly intertwined with a vast causal network of many other things, and so no single thing can ever truly be thought of as independent, as having its own essence. This is the fundamental Buddhist truth of what is called *dependent co-origination.*

"Now, here comes the intriguing connection to sex. Shunyata would seem to be the fundamental openness of reality, its receptivity, the yielding framework through which and within which change and fluctuation and movement become possible. It is as if we are saying, à la Nagarjuna, that it is only because of zero (Shunyata) that there can be variation in intensity (number). Without emptiness there could be no movement; without zero, there could be no numbers. Does one dare to hazard the (by now obvious) surmise that zero is the (in a sense!) principle of the womb, the vagina, and that the numbers, that is to say numerical quantities as opposed to zero, are the principle of the phallus? Are enumeration, measurement, even the ticking off of a Geiger counter or digital display perhaps, an echo of . . . sexual intercourse, where numbers move back and forth in a field opened up to their waxing and waning only by the blessing of a receptive, enveloping vacuity ready to receive them?"

Jacob's theory was fascinating and intriguing, and I looked forward to pursuing these ideas further.

19

TO WHILE AWAY MY TIME WHILE WAITING FOR INFOR-
mation on the Khmer inscription with the first zero, I went to
Jim Thompson's house, one of my favorite sites in Bangkok. Jim
Thompson, born in 1906, was an American businessman, graduate of Princeton, and CIA operative during World War II. He then
abandoned a successful business career in New York to come to live
in Thailand.

Here, he made one of the most remarkable contributions a for-
eigner has ever made to the country. He single-handedly revived a
dying cottage industry: silk production. Within a few years, his vision and business acumen turned Thailand into a major world pro-
ducer of silk and silk products. He did it by providing incentives to
small manufacturers all over Thailand, mom-and-pop businesses,
to weave silk, and he arranged for its sale on fair terms to export
companies.

Thompson became a prominent expat living in Bangkok, and
it is likely that he knew another leading expat: George Cœdès.
There were many functions in which leaders in the close-knit expatriate community of the city met and interacted. But we have no

clear evidence that they did indeed meet. Thompson was divorced when he came to Thailand, and here he knew many women in the European and American community; several of them became intimate friends, although he did not have a long-term romantic attachment, as far as we know.

Thompson built a house—actually a series of several connected houses—in the heart of the city by a canal. These buildings were designed in the typical style of the Thai countryside: made from wood, elevated on stilts to prevent flooding from overflowing rivers or canals, and painted dark red. He was also an avid collector of Asian art, and so today his houses, still containing his impressive collection of fine Asian art treasures, function as a museum.

In 1967, when he was 61 years old, Thompson took a trip to neighboring Malaysia with three friends, a couple and a woman friend of his. They went to a forest recreation area called the Cameron Highlands, where they stayed in a lodge. In the late afternoon, Thompson told his friends he was going for a walk and left the compound to follow a hiking trail. He was never seen again.

Within hours of his disappearance, a large search party was organized, including hundreds of police and other public safety personnel scouring the area in search of Thompson. The entire mountainous region was searched methodically for several weeks, as he was a prominent foreign missing person. But to this day, not a single credible clue has surfaced about Jim Thompson's fate. His disappearance is one of the greatest mysteries of this kind.

I came to Jim Thompson's house to ponder disappearances. A somewhat similar story is that of the brilliant Italian theoretical physicist Ettore Majorana, who had worked with Enrico

Fermi in Rome. In 1938, Majorana took a ferry from Sicily to Naples, where he was living at that time, and disappeared without leaving a trace. As in the case of Thompson, conjectures and theories abound about what might have happened to him. One hypothesis was that he did make it to shore but left the ferry unseen and then went into a monastery to hide from the world, perhaps sensing that a terrible war was about to erupt and that his and Fermi's work in physics might be used to make a doomsday weapon.

I also thought about yet another vanished person, one whose work is so close to my topic: Alexander Grothendieck. We have good indications that he is, indeed, alive. Majorana and Thompson never left behind evidence that they were still living—but who knows, maybe either or both lived for at least some time after their disappearances.

We know that Grothendieck is still alive because he does send communiqués from time to time. The last one was in 2011, when he sent a letter from his hiding place, addressed to a Paris mathematician, in which he demanded that all his published and unpublished works be immediately pulled from any kind of circulation, private or public or anything in-between. Surprisingly, his colleagues agreed to this demand, even though it meant that the mathematics world would lose access to his work. Within days, most of his publications—even copies existing in cyberspace—were removed from circulation. Fortunately for me, I had already secured a copy of Grothendieck's most bizarre, and gargantuan (stretching over 929 pages), mathematical-autobiographical screed titled *Recoltes et Semailles* (Reapings and sowings). This rambling document, written in French and circulated among his friends in manuscript form in 1986, is a mixture of mathematics, biographical descriptions, and thoughts about the universe. He had hoped to get it published but had not succeeded in doing so. In the meantime, the manuscript had achieved great success among mathematicians, although most of its copies, paper or electronic, had by now been destroyed in compliance with his request.

Fittingly, I now sat under a banyan tree in the garden of another missing person and read from my copy of Grothendieck's book. Grothendieck describes how he had been fascinated with the idea of numbers since he was a very young child still living in Hamburg, cared for by a foster family as his anarchist parents, Hanka Grothendieck and Sacha Schapiro (they never married; Grothendieck uses his mother's last name), were fighting with the Republicans on the losing side in the Spanish Civil War of 1936. When the Republicans were defeated by Franco's fascists and routed out of Spain, the couple recrossed the Pyrenees into France but were immediately caught by the French police. They eventually ended up in detention camps. Alexander would join his mother to spend the Second World War in a wartime camp, while his father was sent to his death at Auschwitz.

Numbers, to Grothendieck, were everything. The magic of a number—this most powerful invention of the mind, or discovery of a preexisting truth—was astounding, and he could not stop thinking about numbers and how they came to be. On page 31 of *Recoltes et Semailles,* he writes that when he was small, he loved going to school. (In a French camp for "undesirables," where he lived with his mother, going to school was a rare privilege.) At school, he wrote, "Il avait la magie des nombres" (There was the magic of numbers).

But the world also has shapes and forms and geometry and measure, as the child knew, and while he was still a schoolboy he dreamed to complete the goal the ancient Greeks had pursued, of uniting numbers with shapes and geometry. On page 48, Grothendieck already writes about what he describes as: "Les épousailles du nombre et de la grandeur" (The wedding of number and size). It would be this idea—starting with the invention of numbers—that would lead him to his greatest achievements, including inventing revolutionary concepts such as the entities he called a *motive,* a *sheaf,* and the *topos.* All these abstract concepts derive from the basic idea of a number but extend it to immensely vast realms of highly theoretical mathematics.

Algebra was linked with geometry through algebraic geometry—a field in which Grothendieck would leave his greatest mark. Geometry, the theory of shapes, is extended to topology, an area of mathematics concerned with notions of shape in a more abstract way: deformations of spaces by continuous functions and ideas of distance. It was in this area that Grothendieck defined sheaves and the topos.

Prime numbers were important for Grothendieck's work—as they are to most mathematicians since they are the building blocks of numbers (the nonprimes are made up of products of the prime numbers, hence the primes are elemental). At one time, Grothendieck was giving a lecture on some topic in which he used prime numbers as the skeleton on which to flesh out his general results.

A member of the audience raised his hand and asked, "Can you please give us a concrete example?" Grothendieck said, "You mean an actual prime number?" The questioner said yes. Grothendieck was impatient to continue with his highly abstract derivation, and so, unthinkingly, he said, "Fine, take 57," and went back to the board. Of course 57 is not a prime number, as it is equal to 19 times 3, so this number has become affectionately known as "Grothendieck's prime."

Grothendieck, whose work in category theory and the topos would free us from the confines of the theory of sets, also understood that sets and their memberships were the most beautiful way to define numbers in the first place. This highly theoretical definition of what a number actually *is* uses the most powerful idea humans have ever come up with—that of a complete emptiness, the void, Shunyata. In mathematics, absolute nothingness is defined as the empty set.

And it turns out that we can define the numbers—using the empty set—as follows: Zero is simply the empty set; we now define the number 1 as *the set whose only member is the empty set*. We can now define 2 as the set that contains two distinct elements: the empty set and the set containing the empty set. The number 3 will be a set that contains the empty set, the set containing the empty set, and the set comprising the empty set and the set containing the empty set. Continuing in this way, having started with sheer emptiness and the idea of a set, we can define all the natural numbers (the positive integers) all the way to infinity. As we see, each number is contained within the next-larger number as a series of Russian dolls each placed inside its larger mother. It was this derivation that I thought about when Jacob explained his Shunyata-womb idea. In a sense, the empty set here "gives birth" to all the numbers.

From such concepts Grothendieck, the great master, was able to construct very complicated mathematics. But did he really know about the Eastern concept of nothingness—the Buddhist Shunyata? Well, for much of his life Grothendieck was indeed a Buddhist. And even when he wasn't, he followed Buddhist ideas of peacefulness, charity to others, and dietary habits. He founded an antiwar survivalist group called Survivre Pour Vivre (Survive to live); his home was always open to people who were destitute and needed help; and he was active in many antiwar and environmental groups.

During the great student demonstrations in Paris in 1968—the year he turned 40 and saw it as a milestone—Grothendieck decided to abandon mathematics (although he did produce some mathematical work over the years to come). I wondered if the zero and the Eastern void perhaps played important roles in the life of this leading mathematician. How much did Buddhism influence the thinking of Alexander Grothendieck? I didn't know the answer.

20

WHEN I CAME BACK TO MY HOTEL AND TURNED ON MY
computer, there was a much-anticipated message from Rotanak
Yang (Andy Brouwer's friend). He told me that his father, who
spoke no English, was the director of an institution called Angkor
Conservation, where many Cambodian inscriptions, statuary, and
other artifacts have been taken over the decades to protect and
conserve them. K-127 might have been placed there. But Angkor
Conservation had been looted in 1990 by the Khmer Rouge in
the last recrudescence of their violence in Cambodia, and many of
the pieces kept there had been destroyed or plundered. So it was
unclear whether K-127 would still be there, if indeed it had ever
been brought to this repository. And then he wrote, "But neither
I nor my father can help you any further. In order for us to give
you any more information, you need to contact the Cambodian
Ministry of Culture and Fine Arts to obtain permission to receive
our information."

I closed my PC and sighed. Here I was, in Bangkok, waiting
to go to Cambodia to look for the missing inscription, and now I
needed to deal with a bureaucracy I didn't know or understand.

I thought about this new hurdle and sent a message back to Rotanak: "Would you please give me some idea as to where I should start this request? Do you know anyone at the Ministry of Culture and Fine Arts who I could contact?" I sent the message and went into town to visit Buddhist temples for inspiration.

Next to my hotel was one of the stops for the boats that ferry tourists and Thais up and down the Chao Phraya. I caught one of these packed boats going north and disembarked close to the Royal Palace. The palace was closed—as it was a holiday, the king's birthday—and an armed guard chased me away when I tried to walk in through one of its gates. I crossed the street and within a block found the largest of the temple complexes in Bangkok, Wat Pho. This was the home of the famous golden reclining Buddha. I admired the 150-foot-long gold statue of the Buddha lying on his side, supporting his head with his hand. A sign inside the temple read, "Beware Non-Thai Pickpocket Gangs." I instinctively touched my pocket—my wallet was still there.

Outside, as I crossed the crowded street heading in the direction of the boat landing, a middle-aged man rushed toward me. He opened a color-picture brochure displaying photos of a naked woman. "Young girl," he said, "young girl." I literally pushed him away. This was the scourge of Southeast Asia. Ever since the Vietnam War, when the US military would bring its war-tired soldiers for rest and recreation in Bangkok, the people here found a lucrative trade in selling their women and girls to Westerners. But the improving economy has made the phenomenon less prevalent in recent years.

A few months earlier, at an international conference in Mexico where I had been invited to speak, I met Nicholas Kristof of the *New York Times*. When I told him that I was headed to Cambodia, he said that he had just returned from there. "When I was there, I bought from a brothel two Cambodian girls who had been forced into prostitution, and set them free and sent them into a program that would train them to live on their own," he said. I was glad the world had people like him. He probably could have also bought the freedom of the young girl this pimp was peddling.

When I returned to the Shangri-La, there was an answer from Rotanak. "Try to contact H. E. Hab Touch," he wrote. "I don't have a phone number or e-mail address, but maybe you can find it." I didn't know what *H. E.* stood for, but I spent some time on the computer looking for Hab Touch. I recognized the name as that of one of the former directors of the Cambodian National Museum in Phnom Penh, which was a good sign. This man must know much about antiquities, I felt, and I hoped he would sup port my quest. I located his address and wrote him an e-mail message requesting his help in my search. But for a while there was no response.

After a few days, to my delight, Hab Touch answered my e-mail. He told me he and his people would look into it and try to find more information about the inscription's whereabouts. Mr. Hab (in Cambodia the surname comes first) then indeed spent a significant amount of time trying to help me find the inscription. He finally determined that on November 22, 1969, the artifact was sent to the place called Angkor Conservation (where Rotanak's

father worked) in Siem Reap, home of Angkor Wat and a thousand
other smaller temples in the jungles and fields of western Cambodia. What happened to it afterward, no one knew. He suggested
that I also contact the director of the local museum. It turned out
that Chamroeun Chhan, whom my friend the art dealer in Bangkok had suggested I contact, was the director of that local museum
in Siem Reap, and I was glad that two leads now pointed me to
him. I would try to contact him later, but my greatest need now was
to obtain information about K-127 from Angkor Conservation.

I therefore returned to Rotanak, but he insisted that I get formal permission to be given any more information on artifacts at
Angkor Conservation. So again I wrote to Hab Touch, and in a
few days the desired permission arrived. I was cleared to travel to
Siem Reap to visit Angkor Conservation to look for the lost artifact. I could not believe my search was about to begin in earnest,
and with a specific destination. I took note of what I now knew:
K-127 was discovered in 1891 at Sambor on Mekong. By 1931,
George Cœdès had translated it and realized that it contained the
oldest extant zero and published his finding. The stele was taken to
the national museum in Phnom Penh. In November 1969, it was
moved to Angkor Conservation in Siem Reap. In 1990, a thousand artifacts from Angkor Conservation were destroyed or stolen
by the Khmer Rouge. Now I had official permission to search for
it at its last-known home. I packed carefully, planning for what
could be a difficult search for something that might or might not
be found. This quest could take a long time.

21

I WAS NOW HEADED TO SIEM REAP TO SEE IF K-127 STILL
existed and could be located. Most important in this search, Hab
Touch had told me on the phone that he would get me in touch
with his people at Angkor Conservation, to see if they could locate
this elusive artifact.

With these hopeful signs, I took a long taxi ride to the Don
Meuang airport, the secondary Bangkok airport, and boarded
an Air Asia flight to Siem Reap. Air Asia provided a comfortable
flight in a two-engine turboprop plane, which can use far smaller
runways than jets do. It took off after a couple of minutes on the
runway, but the airline provided no food or drink—you had to pay
even for a glass of water. On arrival, I went through the lengthy
visa application process, had my picture taken, paid the fee (I knew
they only accepted US dollars and I was prepared), and after an
hour of immigration bureaucracy I emerged to find a cab to take
me to my hotel.

The Angkor Miracle Resort Hotel, which caters mostly to
Chinese visitors, turned out to be comfortable and surprisingly
quiet given the number of tour buses that arrive each day. It was

January—high season here—and tourists mobbed the city. When I asked for a taxi, I got the only driver in town who knew absolutely no English or French: not *hello,* not *yes* or *no.* The reason for this was that he was not really a cab driver. The hotel's concierge had called the last available taxi driver in town for me. But even this man was now too busy driving tourists around; so he volunteered his father to drive me. I had never before had to rely on someone who could not even guess what *yes* or *no* meant (and head motions don't work here, since in Asia they often mean different—even opposite—things from what they mean in our culture).

Angkor Conservation, a small, specialized foundation dedicated to the conservation of Cambodian antiquities, is not indicated on any tourist map. So I knew it would be hard to find it under any circumstances, and nearly impossible with a driver who couldn't communicate with me. Fortunately, the concierge at the hotel thought to use his iPhone to find Angkor Conservation. He circled its location on the map he handed me. It was situated by the Siem Reap River, away from the main part of town, and the map showed that the general place he had circled was near an Italian restaurant named Ciao.

I knew this would be hard. I showed the cab driver the map and he grumbled something in Khmer. After a few moments of talking past each other, he decided to move and drove somewhere—I wasn't sure he understood anything. Siem Reap is still a traditional Southeast Asian town, in which cars are relatively rare and most locals use bicycles or, if they are wealthier, motorcycles. Transportation for hire is usually by tuk-tuk. We managed to maneuver our way through a series of traffic jams in which motorcycles and tuk-tuks competed for road space and ignored the legal direction of traffic.

We made our way through the downtown area and headed north on Charles de Gaulle Boulevard, in the direction of the Angkor Wat complex. After passing the lush tropical gardens belonging to a new hotel, we turned right on a narrow road, once paved but now gravelly and lined with rickety tables set up by scores of vendors hawking fruits and vegetables. We continued toward the river. Then the driver turned right into a dirt road and we bumped along for some time, finally stopping by a small farm with chickens running around in the dusty yard, some of them scattering at the approach of our car. There was nobody around. We both got out and stood there, scratching our heads and staring blankly at a map that seemed to say nothing to either of us.

Some ten minutes later, someone emerged from a shack some distance away and approached us. The driver and the man who came over started an animated discussion apparently about where we were and where we were going; both men soon became very excited, raising their hands up in the air, pointing first in one direction, then in another, their voices rising. Finally, the driver returned to the car and slammed the driver's door. I tried to say, "This can't be right. This is not the place . . ." But he understood nothing, and seemed to care even less. He put the car in reverse and drove backward on the narrow dirt path until we came back to the road. There he just stopped and started talking fast in Khmer, looking at me expectantly.

This is crazy, I thought. I picked up the map. I couldn't make out anything—and there was no Ciao restaurant anywhere, either.

After trying to find the place while driving slowly back along the road for another half an hour, up and down and scrutinizing both sides, I decided it was time to give up. "Hotel!" I said. He understood. "Hotel," he said, smiling for the first time since I hired him, "hotel," and he made the way back double-time, traffic and all.

It was late afternoon, and as I got out of the cab at the Angkor Miracle Resort Hotel's driveway, I noticed five tuk-tuks parked at the entrance, their drivers lazing in the sun by their vehicles. One of them, a boy with a soft, round face who looked 16 or even younger, noticed me and ran over. "Sir," he said, "please, please hire me; I am a good driver, and I need the money. Please." The tuk-tuk behind him had in big letters on its side "Mr. Bee." I dislike tuk-tuks: I find them unsafe since the bed behind the motorcycle engine has no protection at all, generally being made of wood, and the ride is bumpy.

He looked up at me. He was small and thin, his eyes bright and pleading. "Sure, Mr. Bee," I said. "But the place I am looking for is closed by now." He looked very disappointed, and his head dropped somewhat; he started to turn to walk away. "Tomorrow," I said. "I promise, really!"—I knew that these tuk-tuk drivers hear *tomorrow* and know there likely will be nothing: The customer will find something else to do, or another driver. "Be here at eight in the morning, please, and I will hire you for a full day—I promise."

Mr. Bee smiled. "I'll be here," he said. I gave him $2 to show my good faith.

THE NEXT MORNING before eight, Mr. Bee was there. He saw me from afar as I approached the hotel door and ran over. "Good morning, sir," he said.

"Good to see you, Mr. Bee," I answered. Then I showed him the map. "Can you read it?"

"Yes, I can," he said confidently and we looked it over together. He smiled and said, "I'll take you there," and he put on his helmet—few other tuk-tuk drivers use helmets, and I took this as a sign of his being very careful; he would prove himself to be careful, intelligent, and considerate. He helped me into the tuk-tuk's cab. It became clear very quickly that not only did he speak close to perfect English, but this young man, a boy, really, was very bright—far cleverer than the driver from the day before.

"First, let's go to the Siem Reap Museum," I said. "And then we'll try to find what I need." We drove through heavy traffic until we reached the museum. I asked Mr. Bee to wait for a little while outside and went in. "I need to see Mr. Chamroeun Chhan," I said, "the director." The woman at the desk asked what it was about. "Mr. Hab Touch sent me to speak with the director about an inscription I am looking for." She picked up the phone and spoke in Khmer. The only words I understood were "His Excellency Hab Touch." And then it dawned on me: I had been dealing all along with perhaps the highest official responsible for antiquities in all Cambodia—someone accorded such a lofty honorific. And I realized how ignorant I had been in not understanding what *H. E.* meant in Rotanak's e-mail message about "H. E. Hab Touch." I was embarrassed that I had not used the proper salutations when communicating with someone who had been so helpful in my quest. Surely His Excellency had many more important things to do than help a random academic looking for an obscure inscription.

Mr. Chamroeun Chhan, on the phone with the person at the desk, soon understood what I was looking for and who had sent me there, but he was away till late afternoon. He said that I should go directly to Angkor Conservation, as Hab Touch had suggested, and see what I could find there. If not, he would meet me at the museum at 5 p.m. I thanked the woman at the desk and went outside the museum and climbed back into Mr. Bee's tuk-tuk.

Mr. Bee made the drive pleasant—he was able to find the shortest route from the museum to the general area we were going to, east of Charles de Gaulle Boulevard in the vicinity of the pediatric hospital, on the road north from downtown. When we reached the hospital, he signaled me that he was stopping.

"May I see the map again?" he asked. I gave it to him. "We turn here," he said, "not by Sofitel as you did yesterday." Sure of what he was doing, he took a barely paved road that ran just parallel to the one we had taken the day before. They were very close and looked the same, with the same street hawkers grazing the sides of each, and the same low bushes by the sides of the road and the same palm trees a little distance away. But the entrance to this road was hidden from the boulevard, and it took a keen eye to discern it.

Mr. Bee signaled with his hand and carefully turned right. He drove slowly, looking searchingly to both sides. About half a mile later, he slowed down. On the left side there was an iron gate with a small sign: "Angkor Conservation." He blew his horn lightly, and a man walked over and opened the gate just wide enough for a tuk-tuk to enter. We went in, driving over a series of stone slabs. A hundred feet inside were several large sheds and a group of men sitting outside one of them, making coffee in a copper pot on a little open wood fire. Mr. Bee approached them.

There was a lot of talking in Khmer and gesturing, but the men didn't understand what we wanted. Mr. Bee smiled sheepishly as he came back to me, and raised his hands up in the air as if to say, "I don't get it."

"Ask them where the director is," I said. There was more gesturing and words exchanged. Finally one of the workers pointed left. We walked in that direction. About 80 feet ahead of us was a makeshift office. I entered; Bee waited outside. "I am looking for the director," I said to the woman at the desk. She ignored me. She clearly didn't understand a word. It was very hot inside, the sun beating down on the tin roof even at this time of morning. I stood and waited. Finally, a middle-aged man came in. "Hello," I said. "His Excellency Hab Touch of the Ministry of Culture sent me here. I am looking for inscription K-127."

"Hello," he said. "Yes, he told us you would be coming. Let me take you to where all the old artifacts are, and you can look for it yourself, if it's there at all. You know what happened right here in 1990 . . ."

"Yes," I said sadly. "The Khmer Rouge came and destroyed much of what you had here. But I hope . . ." He gave a wan smile and motioned me to follow him.

We walked over to a large shed covered in plastic sheeting. "It might be here, somewhere, unless they got it when they plundered this place. You can see what they did." He pointed toward the edge of the enclosure, where there was a pile of broken stones that once represented statues. It looked depressing and hopeless. Then he just turned around and went back to his office, leaving me among the rubble and the many standing artifacts.

I looked around me, away from the piles of destroyed statues. There were probably thousands of items, including heads of statues from Angkor Wat—many hundreds of them, more than half of them badly damaged—and large stones with inscriptions filling much of the area. I began to walk slowly, from one item to the next, inspecting each. I knew vaguely what K-127 had to look like—if it still existed and was in one piece. It would have been a slab of red stone, about five feet high and broken at the top. I walked around for an hour, finding nothing. This was frustrating, and I was losing hope.

Then I just sat down on one of the stone inscriptions, wiped my forehead—it was intensely hot and humid in this deserted, airless shed—and took a drink from my bottle of water. The temperature was surely over 100 degrees. I felt limp and lethargic. Then I forced myself to get up and started examining the stones again. I thought my random search would probably not be effective and decided to look more systematically, row by row. I went back to the entrance and took the first row, going down to the end, and then continuing to the next row, and so on. I searched in this way for another hour, but found nothing.

Finally, I decided to walk *behind* the artifacts. I moved slowly from item to item until I came to a stone that fit the description I had with me. On its back side I saw an old piece of tape pasted to the bottom of the red stone. It read, "K-127." I could not believe my eyes. Could this be right? Had I really found it?

I looked at the front part of this large reddish stone, and there it was—I recognized the Khmer numerals: 605. The zero was a dot—the first known zero. Was this really it? I read it again. The inscription was remarkably clear. I stood by it feeling euphoric. I wanted to touch it, but dared not. It was a solid piece of carved stone, which had withstood the ravages of 13 centuries and was still as legible and clear and shiny of surface as ever. But I viewed it as fragile and delicate; I felt it was so precious that I dare not breathe on it. Perhaps this was a dream, and if I touched the stone, it would disappear. I'd worked so hard to find it. *This is the Holy Grail of all of mathematics,* I thought. *And I found it.*

I really didn't know what to do. I had no plans for how to proceed. I just stood there, amid hundreds of discarded, old, mostly broken sandstone heads from Angkor Wat, steles and pieces of carved stone, fragments of inscriptions of all kinds—and K-127.

The moment of discovery of K-127, the seventh-century stele bearing the earliest known zero of our system.

Its history flashed before my eyes. I imagined it being discovered in the nineteenth century among the ruins of Sambor on the Mekong, in the jungle by the mighty river. I saw it brought to the lab of George Cœdès; I imagined how elated the scholar must have been when he realized he had in front of him a zero carved on a stone in 683 CE, positive proof that would defeat any hypothesis that the zero was a European or an Arab invention. I imagined Cœdès frantically writing his article, making a pencil rubbing of the precious digits 6–0–5, and triumphantly delivering his coup de grâce to his bitter, prejudiced academic opponent. The very real red inscribed stone in front of me provided all that Cœdès needed. This was it—the deciding zero.

I imagined what happened later. The stele was discarded, placed in a desolate museum no one visited, moved to some unknown storage place among other abandoned objects, escaped destruction as Khmer Rouge thugs came rummaging through, smashing and burning and destroying anything of historical or artistic value. I saw it remain unnoticed and unharmed as destruction was perpetrated all around it; then it was moved once again, to come to an ignominious end in this shed in the middle of an empty field tucked in a jungle clearing. I closed my eyes and relaxed from the tension. Here it was, again. K-127. I looked very carefully at the ancient numerals.

After four years of hard work, I had found the Cœdès inscription. It took so much detective work, writing, calling, pleading; so much help from the highest levels of the Cambodian government, including the director of cultural affairs in the Ministry of Culture and Fine Arts; and the assistance of the Sloan Foundation, whose

The face of the stele K-127, showing the ancient writings, including the number 605.

The numerals 605, the zero being the dot in the middle, from the photograph above.

The upper part of artifact K-127, showing its broken top. The numerals 605 are on the second line from the bottom.

The author with K-127, right after verifying the discovery.

kind, intelligent administrators were willing to help a researcher recapture a piece of our lost history. I was euphoric. My quest was over. I would now photograph the inscription, go home, and write down all that had happened to bring me to this moment of final triumph.

Yes, I finally found this red stone with the first zero ever made—still the first known zero of our system, after all these decades, to the best of our knowledge. I remained standing quietly next to the inscription and took many photographs of it.

And then I made my worst mistake.

22

berg's iconic film *Raiders of the Lost Ark* in which Indiana Jones, played by Harrison Ford, goes through a harrowing set of experiences in a South American jungle, complete with skeletons on springs that come to life when a wire is tripped and shoot poison arrows at the intruder who dared penetrate their treasure's hideaway. Then despite the dangers and odds against him, he finds his prize, a golden idol. But just as he comes out of this great adventure, there is his arch enemy, the Nazi-collaborating French archaeologist René Belloq, who with pointed guns takes away the prize, saying, "Once again, Dr. Jones, something that was briefly yours is now mine!" He laughs as he hauls away the priceless archaeological find.

But Jones was smart and shrewd—he was just temporarily outmaneuvered by his foe. I, on the other hand, acted stupidly. Had I said nothing and just walked away, my story would end here. But, like the proverbial crow that couldn't keep his mouth shut and lost the cheese to the wily fox, I opened my mouth. While Indiana Jones's nemesis was a male French archaeologist, mine would turn out to be a female Italian archaeologist—a Sicilian from Palermo.

While I was still gazing in awe and wonder at K-127, this re-markable inscription that was the culmination of a quest that had brought me halfway around the world, a pair of women researchers dressed in lab coats walked into this—rarely ever entered—shed in the middle of a field in this nearly deserted compound in the wilds of Cambodia. They were chatting loudly in Italian.

Fond memories of my parents and my childhood sailings on the ship surfaced. Our ship often called at Italian ports, and I both speak well and love Italian. My father had studied navigation in Trieste, Italy, before becoming a captain, and my mother had taught herself Italian and was the president of the Haifa chapter of the Italian cultural society Dante Alighieri. I longed to speak it again, and in Asia I had little opportunity to do so. And the ela-tion of having rediscovered K-127 made me more talkative than ever—I had a need, a desire to share my exciting discovery with other people.

So I walked over to the two women and struck up a conversa-tion; they seemed pleased to speak to someone who knew their own language. They introduced themselves as Lorella Pellegrino and Francesca Taormina. Taormina was blond and thin. Pellegrino was large, nearly six feet, with a round face framed with unruly, curly, and very dark hair.

I proudly pointed to the red inscribed stone and said, "This is K-127—it bears the oldest known zero in history." They looked at it casually, and both said, "Really?" Then Pellegrino said, "How won-derful! Thank you so much for informing us about this." She turned to face it, and with a small stick she was carrying casually hit the top of the stone, where the writing had been broken off centuries ago.

It made a loud thud. I almost fainted—how could she be so crude and careless, slamming a hard stick against such a precious object? She seemed oblivious to my instinctive recoil and said, "You see, Francesca and I just came inside here to pick up one stone object at random to restore—as practice for our archaeology students—so now we will pick this one, since you told us that it's so important."

I had a sinking feeling in my stomach. *What have I done?* I asked myself. These two Italians were grabbing *my* find—right in front of me—after so many years of my searching for it. Did they even fully understand its meaning to science history? They explained to me that they were both here by agreement between the Italian and Cambodian governments. Cambodia needed archaeologists to work on the vast ancient treasures found in this history-rich land, to learn to clean, catalog, and prepare artifacts for display in museums and exhibits. The two Sicilian archaeologists from the University of Palermo were training Cambodian students at a site near Angkor Conservation so that someday these students would become accomplished archaeologists and help preserve their nation's heritage.

Pellegrino and Taormina, just by chance, decided this day to come to the Angkor Conservation shed to select a random discarded artifact on which to practice archaeological "restoration" with their students. I tried to protest, saying that K-127 was an invaluable object of immense historical importance and that it should not be used as a guinea pig for students to make their mistakes with, spill paint on, or scratch the surface that had withstood the test of time, or perhaps drill into it or grind the facade. They ignored my protests.

"So it's a famous object," Pellegrino said with a self-satisfied smile. "Good! I want the students to experience working with a valuable artifact, not just some old worthless stone. Thank you so much for pointing it out to us." Before I could reply, the two women left the shed to make arrangements to move the heavy stone to their laboratory. I felt faint and powerless.

What could I do? Pellegrino and Taormina were in Siem Reap at the invitation of the Cambodian government. None of the Cambodians would dare oppose their decisions—they were respected professors from a faraway land and had been brought to teach the Cambodians sophisticated technical skills. And I was a visitor, given limited permission to enter the compound to look for a particular item—who was I to interfere with their work? I didn't own K-127—far from it. By intergovernmental agreement, these two Italian professors controlled the artifacts found in this entire area. They could do as they pleased, and they had decided that K-127 was to be used as a learning tool in their lab. They owed no one an explanation.

I didn't want this piece touched—and now they wanted to "restore" it? Restore *what?* It looked perfect as it was. The 1,330 years since it had been chiseled had left it remarkably free of signs of wear; a top segment had already broken off by the time it was found late in the nineteenth century, the part Pellegrino had just struck with her stick. Other than that, the stele was incredibly well preserved. Who would imagine restoring it? It seemed to me that Pellegrino and Taormina were teachers—they wanted to demonstrate to students how to restore artifacts. Whatever item they chose, they only restored it to teach their students this craft. But restoring such an artifact would only bring with it the risk of irreversible damage to an intact archaeological piece of evidence for a great human achievement.

Pellegrino and Taormina came back into the shed, smiling. "It will be moved to our lab in a couple of weeks," Taormina said with apparent satisfaction. And then Pellegrino turned toward me and said, "When you will come back to visit us, you will find the inscription beautiful and restored." *My God,* I thought, *who could ever save it from them now?* I was angry and frustrated and tried to quickly think of a solution.

K-127 belonged to the Cambodian people and deserved to be exhibited, as is, in a museum. I did not understand Pellegrino's insistence on taking it into her lab—why would she need such an important artifact to teach restoration work? It made absolutely no sense to me. I began to suspect that she had another motive: Was she planning to try to pass it off as her own discovery, by publishing an academic paper, perhaps?

And there was something else. It was an issue that archaeology has grappled with for a hundred years. Around the turn of the twentieth century, the great English archaeologist Arthur Evans uncovered the palace of the legendary King Minos at Knossos on the north coast of Crete, the palace Homer described as containing the labyrinth in which the Minotaur—half bull, half man—was confined to roam. In 1628 BCE, a volcano eruption on the island of Santorini created a tsunami that destroyed the palace and the royal city, bringing to an abrupt end the entire Minoan civilization.[1]

Evans did something innovative—and controversial. He "restored" the palace of Knossos. Where a statue of a lion once stood, perhaps, he placed a statue found elsewhere; noticing traces of red paint on it, he repainted it with modern red paint. Where he thought some canals had been built during the Minoan age, he reconstructed them. Where a roofed courtyard loomed over the coastline, he attached a new roof. A visit to the site of Knossos is, therefore, quite different from a visit to any other archaeological site in the world. What Evans did at Knossos had not been done before—at least not to the same extent—nor, perhaps, would ever be done again. Visiting Knossos today is seeing Evans's view of what it *might* have been like—or not.

As we strolled away from K-127, Pellegrino turned to me and said, "Ah, since you're here, why don't you come visit our lab? Soon we will place K-127 there and work on it." My legs were shaking as we entered a small, air-conditioned building in which a handful of young Cambodian trainees were restoring fifteenth- and sixteenth-century (thus, post-Angkor) wooden Buddha statues. A middle-aged Italian professor with a trimmed gray beard supervised the work. Pellegrino brought me over to him.

"This is Professor Antonio Rava," she said, introducing us to each other. Rava was from the University of Turin in northern Italy and had the typical accent of that region, which is close to French (he pronounced the *r* as a French person would, in contrast to the rolling way it is pronounced in most other parts of Italy). He spoke perfect English. "You see, there is very little interest in the post-Angkor era," he said, "so we pretty much have it all to ourselves— nobody cares about these wooden Buddhas made after the fall of the Angkor empire in the fourteenth century. So we restore them, make them beautiful." *Sure,* I thought, *and now you'll do whatever*

Professors Antonio Rava and Lorella Pellegrino in their lab near Siem Reap.

you please to one of the most important artifacts ever found in this country, whose authorities have given you carte blanche to do with antiquities as you desire. But I smiled politely and followed him.

We walked through the lab. At one table a student painted with modern gold color a wooden Buddha statue that showed traces of once having been gilded. Two students were trying to attach a new pedestal to yet another statue. They pushed and twisted and hammered. We walked over, and Rava gave them some instructions and help. Finally, he announced loudly "It fits!," and everyone clapped. Where did the pedestal come from? Nobody knew. It had been found in a pile of statue parts that sat outside the lab, Rava informed me. And there was a statue that lacked an arm, and another Cambodian archaeology student was attempting to attach an arm to it—I had to wonder where that arm came from.

This was not even archaeology as Evans would do it. This was a travesty. I knew from talking with conservators that state of the art methods now emphasize stabilizing an artifact to prevent further deterioration from insects or environmental elements while retaining the integrity of the piece and altering it as little as possible. "When you will come back, you will see how beautifully restored and shining will be the K-127," Pellegrino said as my tour of the lab ended and she showed me to the door. I thought I discerned a mischievous smile on her face, as if to say, "Something that was briefly yours is now mine!" Or maybe it was my overactive imagination responding to the great excitement and stress of this long day.

WITH A HEAVY HEART, I left Siem Reap for Bangkok. In an e-mail message I told Bill Casselman of the University of British

Columbia, the mathematician who had first led me to Cœdès, what had happened. He was clearly furious. "What if they take it to Italy?" he fumed. "And from your photos it is clear that the inscription is in perfect condition and has an invaluable patina—if it is destroyed by their work, this priceless find in the history of mathematics will be worthless and lost forever." He thought it was clear that Lorella Pellegrino now wanted the credit for this find. Casselman offered to write to her, so we both wrote e-mail messages to Pellegrino, in the name of historians of mathematics, pleading with her not to disturb the artifact.

After a sleepless night for me, on January 3 Pellegrino relented somewhat and wrote that she would not "restore" the artifact. "However, I will study it thoroughly, and report the results to the world of science," she wrote me in Italian. I took this to mean that she considered the find her own. I was relieved she wouldn't harm it, as she said. Still, we had to do something to save K-127 more permanently. But what? "Maybe she'll forget about it, do her other work," Debra wrote me in an e-mail. I had a feeling that Pellegrino would not stop. She had fallen upon an important find, and I guessed that she would do everything she could to make it hers. Maybe all the restoration teaching stuff was just a smokescreen—an excuse to take possession of the artifact.

23

TWELVE DAYS PASSED. AND THEN, IT APPEARED: LORELLA
Pellegrino had further plans. "New location for K-127" was on the
subject line of an e-mail she sent me on January 15, 2013, and the
message itself announced that she had moved the giant stone inscription into her own lab and would soon start the work of "studying it structurally and bacteriologically." She attached a photo of
a visitor from Italy, some distinguished professor to whom she
showed the inscription, proudly posing next to it. When I sent
Casselman the photo, he wrote me wryly, "Does he even know
what he is looking at?"

And I feared what would happen once the "structural and
bacteriological studies" began. The stone inscription was perfectly
legible and clear. All that needed to be done with it was to place
it in a museum so mathematicians, historians of science, and the
general public could see it. But Pellegrino refused to give up what
she seemed to perceive was a big fish that had just landed serendipitously in her net. She appeared bound to leave her impression,
her imprimatur, on this "precious find," as she called it (using the
Italian word *prezioso* over and over again).

As if to taunt me (or so I imagined), she sent me messages in Italian every couple of days, describing what she was doing with the inscription. "My students and I have just completed a 3-D study of K-127," she wrote. And a day later she wrote again: "Soon I will start a radiological study of K-127; I will let you know." And then another kind of study was announced. With every one of her messages I became more upset, both at her and at myself for having opened my mouth.

But I was equally determined to right this wrong. Back in Bangkok, I went to consult with my new friend Eric Dieu, the art dealer who originally set me on my trail by giving me the name of Chamroeun Chhan. He was lounging in his gallery when I entered, studying one of his precious art books, perhaps estimating the true value of one of his priceless statues. "Ah, bonjour Monsieur Aczel," he said. "How did it go in Cambodia?" I told him about my adventure, about finding the lost K-127, and then I sat down across the desk from him and put my hands over my face. "But I lost it . . . I stupidly, stupidly lost it, by talking about its importance to a Sicilian archaeologist. I think that she must have thought that if it's that important then she wants it for herself."

"Sorry to hear that," he said. "Now it will likely be sold at auction somewhere in the world and disappear into the hands of some anonymous collector."

He sure knew how to cheer someone up. "Do you think there is anything I can do?" I asked.

"No, I don't see how. She senses it is important—probably without really understanding what makes it important," he said. As a northern European, Dieu clearly did not think highly of

Sicilians—archaeologists or not. He made a joke about how the Sicilian Mafia now might get its hands on K-127. I didn't think the situation was as bad as he'd imagined, but I still worried about the artifact and what might happen to it.

I tried to assess the damage: What was Pellegrino after? Would she really try to sell a rock with an inscription weighing several tons? And how would she get it out of Cambodia? This didn't make much sense to me; it was an art dealer's point of view, I understood. It seemed much more likely to me that as a low-level academic (one whose main responsibility was teaching, not research), she was probably interested in making a name for herself—so that she might afterward be promoted—by writing academic papers about K-127 and receiving credit for a find that wasn't really hers. I had spent too many years as a professor at a small New England college not to be fully aware of the prevalence of such motives in the academic world. But why would she be writing me all these e-mail messages about the artifact? This made little sense to me. Did she, somehow, crave my approval for what she had done and was continuing to do? Weird as it sounded, this seemed to me a reasonable hypothesis.

After getting another message in which she said she was now waiting for some Italian expert in material science to look at K-127 and determine its structural stability, I lost my patience. Fearing these tests might really endanger the artifact, I wrote to her: "This piece is important for mathematicians and historians of science. Its discovery is the result of my years-long research. Had I not walked into that shed at that particular time when we met, you would know absolutely nothing about K-127 and its importance." After this, there were no messages from Pellegrino for a week. Casselman wrote me, "Do you think it was wise to write her such a message?" I had to admit that he was probably right. The following week, there was a terse message from Pellegrino: "You are welcome to visit K-127 any time you like. I am here at the lab with the students from 9 to 12 and from 2 to 5, Monday through Friday, from now until May 15." That was all she wrote. I scratched my head. I was baffled by her responses. Structural analyses were not necessarily noninvasive, and at this point I didn't trust that she wouldn't try to "restore" the piece, as she had originally planned. After all, restoration was what her lab, training, and teaching were all about. Worried and disappointed, I flew back home. I never heard a word from Lorella Pellegrino again.

No, I will not give up, I thought. I will spend every penny I have, if necessary, to save K-127 from being destroyed. I began to plan another trip to Cambodia, with the express purpose of meeting with His Excellency Hab Touch of Cambodia's Ministry of Culture and Fine Arts. I had a long e-mail correspondence with Hab Touch about setting a date to meet in Phnom Penh. We finally settled on dinner on Saturday night, March 23, 2013. I flew to Bangkok a few days earlier and checked into the Hansar Hotel downtown. This was a comfortable hotel with friendly staff, and my large room had space for my rapidly growing collection of working materials, but it certainly was not the Shangri-La, right on the river with its large swimming pool set in a tropical garden. The first thing I did was visit Eric Dieu again.

I wanted Eric's advice on how to handle this situation. I found him sitting at his desk in the gallery in a cheerful and relaxed mood. He was elegantly dressed and sported the same expensive gold watch with the movement visible on the face. "What brings you again to Bangkok, professeur?" he asked.

"I am meeting the Cambodian director general of the Department of Cultural Affairs in a few days, in Phnom Penh," I told him, "and I hope I can convince him to wrench K-127 from the Sicilian archaeologist and place it in a museum."

"Well," he said, "I wouldn't hold much hope for that."

"Why?"

"You see, you have to understand how these people think. They have millions of artifacts here, and they just don't mean much to them. There are storerooms full of statues and stone carvings of every kind below the museum floors. Have you seen the Bangkok museum?" I said that I had, and admitted that it was indeed in a very sad shape, with dirt on the floors and paint chipping from the walls. "If you can," Eric said, "try to convince the director general to have K-127 sent to a European or American museum—it would be much more appreciated there."

I thought about this unpleasant possibility for a moment— I believed that the artifact belonged in Cambodia and not elsewhere—and he went on. "These people," he continued, "they only care about money. The only one who wants the inscription, apparently, is that archaeologist of yours from Palermo. And even if she wants it for science's sake—and attaching her name to an important find, as you seem to think—others may easily steal it from her."

"How?" I asked. I had thought K-127 was at least safe from leaving the country, especially since it must weigh several tons.

"Just like that," he said, snapping his fingers. "They just pack it in a crate, send it by train to Thailand, it is loaded onto a ship, and the next thing you know it is in the hands of some collector in Europe or the US. Don't be naive about these things, professeur . . . It's pretty easy to smuggle any kind of artifact, especially when a lot of money is involved—and from your description, this stele could well be worth many millions of dollars because of its historical significance. And, you never know . . . It could be gone already, lost forever."

I told him that it was supposedly still in Siem Reap, since Pellegrino's message said that I could come to see it again any time I wanted. At that moment, a tall blonde woman walked in, and came over to Eric's side. "Votre femme?" I said. "Madame Dieu?" He smiled and nodded, and I introduced myself to her. I now felt it was time for me to leave. "I'll try my best," I told him.

"I hope you save the artifact," he said. "It is a noble cause. I applaud your trying to save it for science, for history . . ."

24

IN THE MORNING OF MARCH 23, 2013, I TOOK THE TRAIN to Bangkok's Suvarnabhumi International Airport and waited to board a Bangkok Airlines flight to Phnom Penh. I pondered the fate of the artifact I was trying to save: Was it still in Cambodia? Could it be saved? Again, I blamed myself: I would not be in this predicament if I hadn't been so talkative.

Some people fish or do crossword puzzles to relax; I think about prime numbers. An inscription so important to the development of numbers sure had an auspicious number itself, 127. It was not only a prime number, but it was also one of the Mersenne prime numbers. Mersenne was a Minim monk living in a monastery in Paris and a close friend of René Descartes. Both loved numbers and mathematics. Mersenne was Descartes's correspondent while the philosopher traveled throughout Europe and was often the only person who knew Descartes's whereabouts. Descartes was secretive and worried about persecution by the Inquisition because of his embrace of the heliocentric theory of Copernicus, which the Catholic Church abhorred. This was about the time of the trial of Galileo, 1633, and many other thinkers and intellectuals worried about their views of nature becoming known to an unfriendly and very powerful religious establishment.

In some of their letters, Descartes and Mersenne discussed numbers. Mersenne became convinced that numbers of the form $2^p - 1$, where p was a prime number, were always prime. Such numbers became known as Mersenne primes. Let's see how it works. The first prime number is 2. So the first Mersenne number is $2^2 - 1 = 3$, which is, indeed, a prime number. So what is the second Mersenne number? Since the second prime number is 3, the second Mersenne number must be: $2^3 - 1 = 7$, which, again, is a prime number. What comes next? $2^5 - 1 = 31$, also a prime. Next comes: $2^7 - 1 = 127$, and indeed this number also happens to be a prime number. So 127 is not only a prime number—it is also the fourth Mersenne prime. It is thus a very special kind of number.

Mersenne thought that what he had was a theorem, meaning that it was *always true* that numbers of this kind are primes. He could not prove this theorem, and in fact it turned out to be false. In fact, the very next Mersenne number is *not* a prime, because: $2^{11} - 1 = 2047$, which is the product of 23 and 89. (But in general, Mersenne numbers are frequently primes.) This makes 127 even more special, I thought, since it is the last Mersenne prime before we reach the first nonprime Mersenne number. I now felt absolutely ready to try to retrieve K-127 for science history, and at that moment my flight was called and I walked fast to the gate.

Standing in line to board the airplane, I remembered the best story I know about a theorem that turned out *not* to be a theorem at all—as in the case of Mersenne's idea on prime numbers.

The Great Internet Mersenne Prime Search

While Mersenne numbers are not always prime, they do provide a very good way of searching for large prime numbers. All one has to do is plug in the largest known prime number, p, to produce the number $2^p - 1$, since this number is a good candidate for being prime. And it will be exponentially larger than p. One can then check, using mathematical routines, whether this new number is indeed prime. A program called GIMPS—the Great Internet Mersenne Prime Search, in which distributed computing worldwide is being used to search for ever larger prime numbers—has resulted in the finding that $2^{57,885,161} - 1$ is a prime number. At this writing, it is the largest known prime.

The well-known Japanese American mathematician Shizuo Kakutani was in Europe right after World War II visiting another mathematician. They took a walk together in the German countryside and discussed mathematics. Now, right after the war, US forces had a large presence everywhere in Germany, with military bases and camps located throughout the western part of the land—the regions that had just been liberated by the Allies. Eventually, the two mathematicians, deep in a discussion, inadvertently walked onto an American military base. The guard at the entrance to the camp ran after them, furiously yelling at them to stop. "What are you doing?" he demanded. "We are talking," answered Kakutani. "Talking about *what?*" shouted the guard. "We are talking about a theorem," answered Kakutani. "*What* theorem?"

the guard demanded. "It doesn't matter," answered Kakutani. "It turned out to be false."[1]

As I stood in line, still waiting to board the Airbus A319 jet on flight PG 933, I closed my eyes and realized that K-127's inscription date, 683 CE, was *also* a prime number. I could not tell how I knew it—it was an intuition. Later, sitting in the airplane, I made a few lengthy mental calculations (I did not bring a calculator and didn't even have pen and paper, so I had to do it all in my head) and verified that 683 was indeed prime. Everything about this search had an almost supernatural mathematical flavor. I hoped the present stage of my quest would be a success, and the fact that 683 was a prime number made me feel that I had found an auspicious sign.

In an e-mail message before I left Bangkok, Casselman had expressed skepticism. He noted that as far as he knew, Cambodia was a country in which corruption was rampant, especially in officialdom, and that I might not get anywhere. I told him about my experience in Laos and added that I was ready to face anything that might happen. A woman sitting next to me on the plane, a Canadian who worked for an international organization in Phnom Penh dedicated to getting women and girls off the streets and into reasonably paying jobs—perhaps the organization to which Nicholas Kristof had taken the two girls—told me some horror stories about her neighbors being threatened with deportation for an overextended visa unless several hundred dollars were paid as a bribe. Someone I told about my trip had asked me if I was willing to bribe an official in order to save K-127. I said absolutely not. Besides, I had a very good feeling about His Excellency Hab Touch, my government connection in the quest to save the inscription.

25

I KNEW THAT I WAS DEALING WITH A VERY MODEST MAN when I called His Excellency Hab Touch from the InterContinental Hotel and he said that he would meet me there and would call my room upon arrival. I said, "No, please let me wait for you downstairs." But he refused: "No, no, you relax in your room and I will call you when I arrive."

Our meeting was set for 6:30 p.m. I had half a day. I sat in my room, high on the ninth floor of the hotel, and looked out the window. The contrast between Phnom Penh and Bangkok was stark. In the Thai capital, every building is shining, bright, and painted white or light blue. They all look like they've just been built; you never see water stains running off windows, the streets are so clean you could walk barefoot on them, and everyone smiles. Thailand is a prosperous land where tourists—a mainstay of the economy—are treated as royal guests. Despite the political problems the country has endured, including martial law, most Thai are immensely proud of their king and his family, and the royal family's pictures adorn many public places.

Traveling from Bangkok to Phnom Penh dramatically illustrates the contrast between a prosperous country and an underdeveloped one. And it really shouldn't be that way. Cambodia deserves much, much better. The view outside my window was drab. The dominant color was not clean white but faded yellow and orange. The city appeared slightly out of focus viewed through a haze of smog or other pollution, and in fact the smell of burning fields or garbage was quite strong and persistent. I took a cab to the river—I had several hours to kill before our meeting. The smaller Tonle Sap River, which starts at the Tonle Sap Lake by Siem Reap, winds its way down to meet the wide and deep Mekong in Phnom Penh. The city's heart lies along these two rivers. The royal palace is there, still with signs of mourning for the recently deceased king, Norodom Sihanouk, who ruled the country through tumultuous times, including having to navigate a very difficult course during the reign of terror by the Khmer Rouge. At that time his powers were suppressed, and he had to walk a fine line between trying to save his people from the brutal massacres and saving his own head so that he could live to fight for his people. Consequently, his legacy is mixed.

I stopped by the famed FCC cafe and bar, on the second floor of a building by the waterfront. FCC stands for Foreign Correspondents Club, and its walls are decorated with many photographs of the courageous journalists who dared expose the horrors of the Khmer Rouge era, from 1975 to 1979, and beyond. These journalists often risked their lives for a photo or a story, and the pictures on the wall bear evidence to their efforts. I had a tropical drink and looked toward the rivers. The streets were thronged with motorcycles and bicycles and the ubiquitous tuk-tuks, their sharp-eyed drivers ever on the lookout for the few foreigners who come to visit this city. I walked around, signaling every few minutes to someone that, no, I did not need a tuk-tuk ride anywhere. The palace grounds were open, but because of the mourning, visits to the insides of the buildings were canceled. Then I returned to my hotel to wait for my visitor.

We had agreed to meet at 6:30, and at 6:15 I got a call from the receptionist that His Excellency was waiting for me. I rushed downstairs. I was surprised that he had come without even a driver or bodyguards. In front of me was a man of medium height with dark hair and a face I recognized from his photo at the national museum, which he had directed for a number of years. He still looked young and energetic. "Let's eat in the hotel," he said. "That would be easiest, I think." We walked over to the hotel restaurant, which had just opened for dinner a few minutes earlier. They were serving a full buffet of Asian and Western food. It looked good to both of us, so we went found a table, sat down, and ordered drinks.

We started talking about the museum and about art. He told me he'd worked with statuary for a long enough time that he could tell if a piece a museum was interested in buying was authentic or a fake. "It's just a feeling you have, a sixth sense," he said. "There is no scientific or other reason for your decision. You look at a statue and something just doesn't look right—it's a sixth sense." That's interesting, I thought: Some people can sense whether or not a number is a prime, and others can tell whether a piece of art is a fake. It's something about how a number looks, or a statue, or—for a police detective—how a suspect may hold himself or act. We seem to often have an extrasensory kind of perception of reality; maybe that's how we invented numbers. He told me how he reached his place in life. "I started out very poor, and starving, and without a place to stay," he said. "You see, I grew up during the Khmer Rouge regime. Until I was 11 years old, I could not attend school."

The Khmer Rouge had separated him from his family—as they did many children—and he was forced to work and was fed very little and often had to sleep on the bare ground. I said I was sorry to hear how bad it was. "But when I finally was allowed to go to school, at 11, being older than most first-graders, I worked extremely hard. And I became the top student in my school. So I won a scholarship to study at university in Poland."

He arrived in Poland, still not more than a boy, with only $80 on him. "It was a lot of money—my family had to work hard for months to save this amount for me. And I had to spend all of it at once to buy a winter coat—it was so freezing cold in Poland. Coming from Cambodia, I never knew what cold was—I had no idea." Then the scholarship money started to come in, and he was able to support himself while studying Polish every day till midnight, and at the same time attending classes in museum conservation and art. He learned how to appreciate art and how to prepare it for display in museums; how to run an exhibit and how to prepare descriptive displays; and how to perform museum administration tasks. After earning his bachelor's degree, followed by a master's and continuing on beyond it, he was offered a teaching position at a Polish university, which he held for two years before returning home to Cambodia. By then his Polish was perfect—but now in Cambodia, he had no opportunity to use it.

In Phnom Penh, he joined the staff of the Cambodian National Museum, working his way up to museum director. Then he accepted the offer of a higher job: the directorship of the Office of Cultural Affairs at the Ministry of Culture and Fine Arts. "I love my job," he said. "I am responsible for all ancient temples in Cambodia—there are over 4,000 of them—and statues and steles and inscriptions of all kinds found anywhere in Cambodia." At that moment, his cell phone rang. "I have to take it," he said. "It's my boss, the Minister."

When he hung up, I remarked that he was using an Android. "Yes," he said. "I love toys like this. Whenever a new device comes out, I need to buy it to play. You see, when I was a child I had not one toy. It was impossible for children to just be children during the time of the Khmer Rouge. You were lucky if you had a meal and a place to sleep on the ground. So now I make up for it and play with toys." I knew that unfortunately this story was still ongoing: in the morning's newspaper I had found at the hotel, the main front-page article was about bringing to trial an octogenarian leader of the Khmer Rouge, almost three and a half decades later.

"Well," I said, "it is about an inscription, a stele from the seventh century, that I need to speak to you." I opened my PC and showed him the photographs of K-127 taken 11 weeks earlier. I explained its immense importance to the history of mathematics and the history of ideas. "This is the first example we know of people using a sign for zero," I said, "other than the Maya zero glyph, which is not connected to our ten numerals. It appears that zero was invented not long before this inscription was made, and it is the earliest known appearance of it." He seemed interested in what I was saying, so I continued. "To me," I said, "K-127 is as important as the Rosetta Stone—or more."

Mr. Hab considered this for a moment and said, "It definitely belongs in our museum."

"Absolutely," I said. "That is exactly why I wanted to see you!—besides thanking you very much for helping me find it. I think that K-127 should be placed in the Cambodian National Museum. I even know where I would put it . . ." I opened a small book I had brought with me, a guide I had bought at the museum, and found a page that displayed a chart explaining the various exhibits. "I think it belongs right here," I said, pointing to a room at the northeast corner of the museum. "This is the area where you currently exhibit seventh-century pre-Angkor statues and steles, and I think it should be placed there."

"Excellent," he said. "Why don't you write the description to go with the display, complete with explanations of its importance and all the references, and I will take care of it."

I was elated. "Thank you, thank you. That is exactly what I have been hoping for. I will send you the description by e-mail within a day or two." We continued our conversation in a relaxed atmosphere. "George Cœdès called your civilization of Angkor and pre-Angkor 'Indianized,' but I don't know if that is true," I said. "To me, calling a civilization in Southeast Asia 'Indianized' is the same as calling American civilization 'Germanized.' Cœdès thought that it was so because the Hindu gods, and Buddha, were worshipped here and Sanskrit was often used. Well, in America we use words in English that come from German, and at Christmas we have Santa Claus. Why would anyone call our civilization

'Germanized,' and by the same token, why should your ancient civilization be considered 'Indianized'? Besides," I said, "K-127 is written in Old Khmer, not Sanskrit."

"Well," he answered, "Old Khmer is derived from Sanskrit, and don't forget that many of the themes you saw in art at Angkor Wat, such as the famous bas relief of the Churning of the Sea of Milk—all these scenes come directly from the Indian epics of the *Ramayana* and the *Mahabharata*. Our civilization was affected mostly by India, and China—another powerful nation in our region—had a lesser influence on us."

"The Chinese called Cambodia Fu-Nan, right?"

"Yes, but that is only how it is called in Chinese records of the time. Others considered Cambodia part of ancient Chenla. And the most important part was called Water Chenla, since water is so important for us—you see, the civilization sprang everywhere that had large amounts of water."

"The Baray?" I asked.

"The Baray and other large sources of water," he said.

"The infinite sea," I said.

"Yes, the infinite sea." I had hoped to be able to argue that the zero was a purely Cambodian, that is, Khmer, invention. But Mr. Hab, who was clearly an expert on his country's art and history and civilization, did see a connection with India early on in the development of Khmer culture. "You see," he continued, "The artistic styles mature and become purely Cambodian only later, after your period of the inscription K-127—this happens during the Angkor period. Before it, there are four or, as I claim there should really be, five distinct artistic styles. But when you reach the very active and prodigious period of Angkor, with its mighty kings eager to forge their own cultural styles, you get many more artistic and architectural ideas, and these are purely Cambodian. But not so in the earlier times." This was new and fascinating to me.

"But of course, there is a continuity from early times onwards," he said. "What the minister of culture and I spoke about a minute ago was a trip we are making together tomorrow. I have to wake up at 5 a.m., and we leave for a long car drive to a site that has been discovered 50 years ago, but whose continuing excavation is now yielding incredibly important results. The ruins there are from 4,000 BC."

"Amazing," I said. "Your civilization started in the Neolithic . . ."

"Yes. There are mostly stone tools there, but they are very interesting and show a high stage of development. Our civilization is indeed very old." I asked him about the later periods. "Well," he said, "many of the temples you find in Cambodia—there are 4,000 of them only if you count groups of temples as one; if you count individual temples, then you have many, many thousands—are from the first few centuries CE, because that's when the religions arrived here: Hinduism and Buddhism, both of Indian origin. And the seventh century is extremely important: That is when you have the temples of Sambor Prei Kuk and Sambor on Mekong built. Maybe the birth of the zero at that time has something to do with the great surge in religious construction." I was pleased to hear this, as it clearly agreed with my belief that in the East numbers, including zero, were originally invented for religious purposes.

"And it continues?" I asked.

"Yes, of course," he said. "There is no break in the civilization. Different kings made their capitals at different locations—always at places where water is plentiful—and so the sites moved. And then in the ninth century there is this move to place the capital at Angkor, and you have a continuous habitation of the site from the Angkor period, with its important kings such as Jayavarman VII, through to the post-Angkor period and today."

This was surprising to me since I had always heard that Henri Mouhot had rediscovered Angkor in the jungle in the 1800s. "So Mouhot didn't really discover anything?"

"No, of course not," he smiled. "That is a Western myth—just like the myth you told me about the numerals being invented in the West, which Cœdès managed to debunk so well using your K-127. There was a continuous settlement at Angkor—people lived all over this area and have lived there for almost a millennium. Mouhot just came in and saw it, and he noted that the temple was somewhat covered by jungle growth—the famous pictures you see of big tree roots encompassing an ancient building. But there were people living everywhere in the vicinity of the temple, and worshipping in it: You know that it now serves as a working Buddhist temple."

"Incredible," I said.

"OK, I need to wake up very early, you know, so I had better go." We said goodbye and I tried to walk him to the hotel's entrance. "No, no, please," he said. "I can find my way fine. Have a good stay, and anything you need, just let me know." I thanked him warmly and promised to send him my write-up to accompany K-127 very soon.

I spent the entire next day in my hotel room working on the description for the display of K-127 to be placed in the museum.

I sent my work to Hab Touch by e-mail and waited for an answer.

Inscription K-127

Discovered in the nineteenth century at Trapang Prei, Site of Sambor On Mekong; seventh century, Pre-Angkor Period.

First Translated from Old Khmer, into French, by George Cœdès and published by him in 1931.

This inscription bears the earliest zero numeral ever discovered.

What is the importance of the zero? Zero is not only a concept of nothingness, which allows us to do arithmetic efficiently, but is also a place-holding device that enables our base-10 number system to work, so that the same 10 numerals can be used at different positions in a number, making our system extremely efficient. The Roman system, for example, which preceded our number system in Europe until the Late Middle Ages, employed Latin letters for quantities (I for 1, X for 10, L for 50, C for 100, M for 1,000); these letters had to be repeated, for example writing MMMCCCLXXIII for the number 3,373. We see that in our system the same numeral 3 is used in three different places, allowing for economy and ease of notation. None of the Latin letters could be repeated in different contexts. In our number system, it is the zero that enables the efficiency and power of the system: Thus, a 5 in the units location is a 5; but the same symbol in the tens location makes it a 50—if we can also use a zero as an empty place-holder for the units. Similarly, 505 can only be written in this highly efficient way because we use zero as a place-holder for the tens. The Babylonian system, for example, which predated the Greco-Roman letter-based number system by about two millennia, used base-60 with no place-holding zero. Hence, the difference between 62 and 3602 (where 3600 is the next-up power of 60) had to be guessed from the context. Our number system, using a much smaller base, and employing a special symbol for zero, derives its immense power and usefulness through the use of this place-holding zero, as compared with the Greco-Roman, Babylonian, and Egyptian systems. When we also consider the fact that arithmetic is so much more powerful with the use of zero, which helps define the whole realm of negative numbers, and the fact that everything we do with a computer (or cellular phone, GPS, or anything electronic) is controlled by strings of zeros and ones, it becomes clear that the invention of zero is perhaps the greatest intellectual achievement of the human mind.

So who invented it?

This inscription, bearing the earliest known zero ever discovered, is written in Old Khmer and begins with the words:

çaka parigraha 605 pankami roc . . .

Translated, it reads:

The çaka era has reached 605 on the fifth day of the waning moon . . .

The zero in the number 605 is the earliest zero of our system we have ever found.

These are Old Khmer numerals for "605," and the dot in the center is a zero—the first zero ever made (as far as our present knowledge goes). The çaka era began in AD 78, so the year of this inscription in the Christian calendar is 605 + 78 = AD 683.

This inscription has a celebrated history. Until the 1930s, many scholars in the West believed that the zero—the key to the efficiency and versatility of our base-10 number system—was either a European or an Arab invention. The oldest known zero was in India, at the Chatur-bhuja temple in the city of Gwalior. That zero is dated to the mid-ninth century. Since that era coincided with extensive Arab trade, it could not be used to defeat the hypothesis that the zero was invented in Europe or in Arabia and from there moved east. The publication of George Cœdès's article in 1931 (see reference below) proved definitively that the zero is an Eastern, and perhaps Cambodian, invention since this inscription predates the Arab empire, as well as the Gwalior zero, by two centuries. It is of note that a zero that is one year younger, thus dating from AD 684, was found at around the same time near Palembang, Indonesia, and was also published by George Cœdès.

Inscription K-127 was kept for a time in this museum but was moved to Angkor Conservation in Siem Reap on November 22, 1969. During the Khmer Rouge regime of terror close to 10,000 artifacts were stolen or defaced and this inscription's whereabouts were uncertain. It was re-discovered in a shed at Angkor Conservation by Professor Amir D. Aczel of Boston University on January 2, 2013, and brought to the attention of His Excellency Hab Touch, whereupon it was reinstated at the museum.

References:

Cœdès, George, "A propos de l'origine des chiffres arabes," *Bulletin of the School of Oriental Studies, University of London*, Vol. 6, No. 2, 1931, pp. 323–328.

Diller, Anthony, "New Zeros and Old Khmer," *The Mon-Khmer Studies Journal,* Vol. 25, 1996, pp. 125–132.

Ifrah, Georges. *The Universal History of Numbers.* New York: Wiley, 2000.

26

I HAD BEEN CONCENTRATING ALL ALONG ON THE IMPOR-
tance of zero as a place-holder within our number system, and on
showing how our numbers work because we are able to insert a sign
that says there are no tens or no hundreds or no thousands, and so
on, in the representation of any number whatsoever using simply
the ten numerals in our decimal number system. But what about
the system as a whole?

Sitting at the departure lounge at the Phnom Penh airport
waiting to board my flight back to Bangkok, I was pondering the
rich history of zero the number—a concept I am convinced could
only have arisen through a purely Eastern way of thinking (and,
independently, by the Maya in the West).

Equally, I was thinking of the idea of infinity, also prevalent
in Eastern thinking: the "endless sea," Ananta the sea serpent, eternity, and innumerable other forms of extent that go beyond the
simple numbers 1, 2, 3, and so on. But the development of our
numbers in a purely mathematical setting, including both Eastern
concepts of zero and infinity, was to take place in the West—or

rather, both in the East and in the West (the rational, irrational, and complex numbers were explored theoretically in Europe between the fifteenth and nineteenth centuries).

We've seen that it is possible to define the numbers starting simply from the void, the empty set, and proceeding through the use of set membership: the set containing the empty set for 1, the set containing the empty set and the set containing the empty set for 2, and so on. But of course this is a sophisticated way of defining the numbers, using sets and the idea of set membership. In reality, numbers developed in a very different way.

The ancient Babylonians and Egyptians, thousands of years ago, learned to assign numbers to objects and thus to abstract the concept of number from the magnitude of the sets of things they observed. Perhaps the greatest intellectual discovery of early antiquity—at the dawn of civilization, really—took place when someone, or very likely several individuals at different places and times, could look at three stones on the ground, three cows in a meadow, three people walking on a path, three grains of wheat, three pyramids, three goats, three children, and understand that all of them had one and only one aspect in common: the quality of "being three." Similarly, four could be defined as that aspect of many different things that are four, and so on. The numbers could grow and grow, and the magic of this understanding—that things that are of the same discrete magnitude are in some sense the same—was, and is, overwhelmingly powerful.

Soon enough, people of antiquity added words to their languages to represent these numbers. In fact, in India especially and in several other Asian countries under its influence, there were special words, nouns that everyone knew belonged to categories that were of universally accepted numbers, and these nouns became synonymous with the numbers. Here is an excellent example, from Cœdès's seminal 1931 article, commenting on the stele of Changal (my translation from the French): "The year of the king of the çaka expressed in numbers is: the flavors, the organs of sense, and the Veda."

Cœdès explained that there were six known flavors of food, five senses, and four Veda (the ancient collections of Hindu holy writings). Thus, this is a way of expressing the number 654 in words. This method was widespread in Cambodia, India, and other countries in south and Southeast Asia.

Next, Cœdès gave the example of an inscription from a place called Dinaya, discovered in 1923, in which the date çaka 682 is given as "the flavors, the Vasu, and the eyes." Again he noted that flavors stood for 6 and explained that Vasu (deities attending to Vishnu, of which there were eight) stood for 8, and we know that a person has two eyes.

But Cœdès also discussed the problems that arose here. Across geographical areas and through time, there was not always complete agreement on which number was represented by which noun; sometimes ambiguities existed.[1] This difficulty is similar to one we encounter today when we use the phonetic alphabet.

When asked to spell my name over the phone to someone whose English may not be perfect, or over a bad telephone connection, I often have to use this words-for-letters system. "Aczel," I say, "A for apple, C for Charlie, Z for Zebra, E for Europe, L for Larry."

I use these words because they are the first ones that pop into my head, and I usually have to repeat this a couple of times.

But of course I am mostly wrong, as the accepted NATO Phonetic Alphabet is: Alfa, Bravo, Charlie, Delta, Echo, Foxtrot, Golf, Hotel, India, Juliet, Kilo, Lima, Mike, November, Oscar, Papa, Quebec, Romeo, Sierra, Tango, Uniform, Victor, Whiskey, Xray, Yankee, Zulu. But does anyone remember these?

By analogy, a system of numbers, which was prevalent in the East for centuries—in which one says "Veda" for four, "flavors" for six, and so on—certainly could not have been uniformly well understood by everyone. This was one important reason why written signs for numbers had to be invented.

Cœdès described the ancient Khmer number system, which was *not* decimal. Even today, despite the borrowing of numbers above 30, which are decimal, lower numbers in modern Khmer are not perfectly based on 10, but also on 5 and on 20. The Khmer, Cœdès noted, use many multiples of 20—as the French do only once, for 80 (quatre vingt, or four twenties) and numbers that go with it (for example, quatre vingt dix neuf, for 99). The Khmer use more of these multiples, clearly a vestige of a base-20 system due to our having ten fingers and ten toes, which persisted. This is reminiscent of the Maya number system, which was almost exclusively base 20 (with the exception of the partial calendrical use of base 18).

In antiquity, Cœdès explained, the Khmer possessed *only* the numbers: 1, 2, 3, 4, 5, 10, 20, and several multiples of 20. This was all they had as far as numbers they understood. At some point, they borrowed the Sanskrit word *chata* for 100. With these numbers they expressed all numerical information.[2] All this was, of course, before the maturing of their numbers and the invention of zero (or its importation from India or some other place) as attested by inscription K-127.

What all this taught me was that fingers and toes are really important. If we had not had them, or had different quantities of them, maybe we would view numbers in a totally different way. If some day we were to meet aliens with only two fingers on each hand and two toes on each foot, their number system might well be binary, allowing them to communicate with the innards of their computers more directly than we do: Our numbers always have to be "translated" into a binary (0 and 1 only) code for a computer to understand them.

On the other hand, with two fingers per hand and two toes per foot, maybe their number system would be octal (based on 8). It was fun to speculate on such things, and it kept me entertained as I waited to hear about the fate of my precious find. In Bangkok, it helped me relieve the immense tension of waiting for news about the fate of K-127 and whether Hab Touch would follow through on his promise.

GEORGE CŒDÈS RETURNED to his native France some years after French colonial rule in Indochina ended, as these new nations grappled with questions of democracy, parliaments, monarchy, and Communism. In Paris, he had a prestigious academic position and continued to write papers and books about Southeast Asia. He was highly decorated, having been awarded the rank of commander in Thailand's Order of the White Elephant, as well as France's prestigious Legion of Honor. He died in Paris in October 1969—a month before K-127 was brought to Angkor Conservation. Cœdès had several children, and one of them became the admiral of the Cambodian fleet. This fact made me feel a kind of nautical kinship with this great man.

27

ON APRIL 9, 2013, I FINALLY GOT THE E-MAIL MESSAGE I had been waiting for:

Dear Professor Amir,

I apologize for having taken so long to write to you. It was a great pleasure to meeting with you in Phnom Penh and delighted to hear about the history of Zero. Thank you for your research article on Khmer Zero, which is now the earliest Zero in the world civilization. I have shared this exciting news with my colleagues and hope this inscription will be on display in the National Museum in Phnom Penh. I look forward to meeting with you again and please let me know if I can be of assistance to you with this important research.

With best wishes,
Touch

I was elated. I couldn't believe a successful conclusion was finally in sight. Could it be that my odyssey was now over? Following e-mails reassured me that what I had hoped for was going to take place at last. His Excellency Hab Touch would arrange for the priceless K-127 to be taken out of the hands of Lorella Pellegrino and placed in the Cambodian National Museum in Phnom Penh, where it once was, and where it belongs. From then on, scholars, mathematicians, historians of science, and the people of Cambodia and elsewhere would be able to see the very first zero of our numbers ever discovered—a find that changed our view of history, the one artifact in the history of science that proved definitively that the zero came from the East.

Debra met me again in Bangkok, and we flew to Paris together a week later, switching planes in Bahrain. At our small hotel on the Left Bank, I used the Internet and wrote a short article about the rediscovery of K-127 for the *Huffington Post*. It was published within hours. After I sent the link to Hab Touch, he responded that he was delighted that people would now be learning about "Khmer Zero," as he called it. "Let the discussion begin!" he wrote me.

His country could benefit from displaying and explaining its antiquities, and I hoped he would succeed in his efforts to repatriate to Cambodia many statues that had been looted during the Khmer Rouge era and sold to museums around the world. The *Tribune* had an article about the New York Metropolitan Museum of Art agreeing to return two such statues, and other museums around the world were considering doing the same. I knew that this was the result of Hab Touch's negotiations with museums, and

I felt good that my own work had contributed in small measure to this larger effort.

Debra left to return to Boston and I stayed in France for a few more days. I had one last thing to do before this big adventure was over. After accompanying her to the counter for her transatlantic crossing, I walked over to another part of Terminal 2 at Charles de Gaulle Airport and boarded a domestic French flight to the south.

2 8

AFTER LANDING IN TOULOUSE, IN SOUTHWEST FRANCE, I
walked over to the desk of a car rental agency and collected the
keys to an Alfa Romeo. I headed due south, to the high Pyrenees.

The Alfa took the twists and turns in the steep mountain road
beautifully. It was exhilarating to drive it uphill through so many
quick, sharp turns. After two hours of climbing, I made it to the
top, way above the tree line, having just crossed the border to the
independent mountaintop principality of Andorra. I enjoyed a
strong espresso at nearly 9,000 feet, was buoyed by the breathtak-
ing view from the summit, and then headed back down somewhat,
recrossing the French border. Two road turns below it, I found
what I had come for.

I stopped by the gate of an alpine villa built of wood, the outside panels carved in the ornate designs one often sees in the Austrian Alps.[1] I knocked on the door, and an attractive woman in
her fifties opened it. She was wearing a long blue dress with a wide
décolletage that revealed generous cleavage. "Oh, he's been waiting
for you all morning," she said with a smile. "Let me get him . . .
Laci!" she called.

He came down the stairs. At 88, he looked fit and healthy. "So good to see you!" he said, giving me a big hug. "It's been so many years . . . what, 40 or so?"

I smiled and said, "Yes, yes, a very long time. But I wanted to see you. And I have something that may interest you." We sat down in the spacious living room that opened to the balcony with its views of the mountains and talked about the old days on the ship, about the mountains, and about mathematics and the birth of the numbers. "You told me something when we parted on the ship so long ago—in 1972," I said. He looked surprised. "The name was George Cœdès," I said, spelling it out. "He was the French archaeologist who found the first known zero in Asia."

"Ah yes," he said. "I vaguely remember something now. So he found it, right."

"But it was lost, you know," I added. "The Khmer Rouge—"

"Ah, yes, they destroyed everything, I've heard. So it is gone now?"

"Well—I did manage to find it," I said.

"Find the first zero?" There was a glint in his old yet still keen eyes.

"Yes. Let me show it to you." I turned on my PC and showed him the photographs of K-127. "This is the oldest zero in history," I said. "I found it after so many years of searching. And it was indeed Cœdès who first published it in 1931, debunking those old claims about the zero being a Western or an Arab invention."

Laci sat on the couch across from me, smiling. "So, my friend," he said, "you found the earliest known zero. Congratulations! That's really something. What will you do next?"

"We still don't know where the numerals as a whole came from. Someone should look into the Indian numbers: Ashoka's, the Nana Ghat's, the Kharosthi. There may be place for good research there, and to see whether, indeed, Aramaic letters have led to the numerals. But as a mathematician, I suppose you aren't interested in this kind of work."

"No," he said, "your idea about the origin of the concept of zero, coming from the Buddhist void, is more interesting to me. Maybe some philosophers will follow with this thread." He paused, and after a moment continued. "But what you've achieved is significant—and I'm so glad that a casual conversation with me so long ago led you to this fruitful search." He was clearly pleased and stood up. "You did it, you did it, I am so proud of you!" He held my hand. "This calls for a drink." He excitedly called for the woman—girlfriend or wife, I never found out—and she brought us whiskey on ice.

Then she opened a jar of black Russian caviar and spread it on little toasts for the three of us. Real Caspian sturgeon—I knew it. It must have cost a pretty penny. "I remember eating caviar on the ship," I said. "But that deep-pocketed shipping company, Zim Passenger Lines, which could afford to decorate the ship's halls with original oils by Chagall and Miró—and then went bankrupt because it had spent so much money on such luxuries—paid for it all."

"Ah, don't worry." Laci straightened up and looked at me. "We get a lot of it here." The woman laughed knowingly, and he walked over from the living room to the adjacent kitchen and opened a large refrigerator, just so I could see what was inside: many more jars of Caspian caviar. And the bar was stocked with a lot of expensive liquor: scotch whisky, Calvados, Drambuie, Grand Marnier, sake. I looked at him a little puzzled.

"Well," he said after a moment, "you saw the French customs checkpoint just up the road, right? You couldn't have missed it." I didn't understand. "You know that Andorra is one of the last tax havens in the world, don't you?" I nodded. A vague notion began to surface in my mind. There was a moment's silence. He looked at me, and then he said, "You know, late at night, there is nobody there at the checkpoint. And this house is at exactly the right place—"

"Just like my mother's suitcase," I interrupted.

The thinnest of smiles spread across those old lips. "Just like your mother's suitcase," he said.

NOTES

CHAPTER 1

1. An analysis of the forms of Latin numerals, and a new theory about their being derived from Etruscan signs, is well presented in Paul Keyser, "The Origin of the Latin Numerals from 1 to 1000," *American Journal of Archaeology* 92 (October 1988): 529–46.

CHAPTER 2

1. Georges Ifrah, *The Universal History of Numbers* (New York: Wiley, 2000) has a number of pictures of ancient bones with markings.
2. Thomas Heath, *A History of Greek Mathematics,* Vol. I (New York: Dover, 1981), 7.
3. For a modern description of this issue, including later contentions by other scholars, see Georges Ifrah, *Universal History of Numbers,* 91.

CHAPTER 3

1. A good description of the Mayan numerals, calendar, and the zero glyph can be found in Charles C. Mann, *1491: New Revelations of the Americas Before Columbus* (New York: Knopf, 2005), 22–23, 242–47.
2. Georges Ifrah, *The Universal History of Numbers* (New York: Wiley, 2000), 360.

CHAPTER 4

1. Saint Augustine, *The City of God* (New York: Modern Library, 2000), 363.

2. Chapter 18, Verse 8 of the *Mulamadhyamakakarika,* written by the prominent Buddhist monk and scholar Nagarjuna in the second century CE.

CHAPTER 5

1. Louise Nicholson, *India* (Washington, DC: National Geographic, 2014), 110.
2. David Eugene Smith, *History of Mathematics, Volume 2: Special Topics in Elementary Mathematics* (Boston: Ginn and Company, 1925), 594.
3. Takao Hayashi has referred me to Alexander Cunningham, "Four Reports Made During the Years 1862–1865," *Archaeological Survey of India* 2 (1871): 434.

CHAPTER 6

1. Mark Zegarelli, *Logic for Dummies* (New York: Wiley, 2007), 20–21.
2. Ibid., 22–23.
3. F. E. J. Linton, "Shedding Some Localic and Linguistic Light on the Tetralemma Conundrums," manuscript, http://tlvp.net/~fej.math.wes/SIPR_AMS-IndiaDoc-MSIE.htm.
4. See Pierre Cartier, "A Mad Day's Work," *Bulletin of the American Mathematical Society* 38, no. 4 (2001): 393.
5. Ibid., 395; italics in the original.
6. F. E. J. Linton, "Shedding Some Localic and Linguistic Light on the Tetralemma Conundrums."
7. More on this can be found in C. K. Raju, "Probability in India," in *Philosophy of Statistics,* Dov Gabbay, Paul Thagard, and John Woods, eds. (San Diego: North Holland, 2011), 1175.

CHAPTER 7

1. Kim Plofker, *Mathematics in India* (Princeton, NJ: Princeton University Press, 2009), 5.
2. John Keay, *India: A History* (New York: Grove Press, 2000), 29.
3. Ibid., 30.
4. Ibid., 30.
5. John McLeish, *The Story of Numbers* (New York: Fawcett Colombine, 1991), 115.
6. Ibid., 116.
7. David Eugene Smith, *History of Mathematics, volume 2: Special Topics in Elementary Mathematics* (Boston: Ginn and Company, 1925), 65.
8. M. E. Aubet, *The Phoenicians and the West* (Cambridge: Cambridge University Press, 2001).

CHAPTER 8

1. Robert Kanigel, *The Man Who Knew Infinity* (New York: Washington Square, 1991), 168.
2. There is a rare reference to this plate, which may have had an early zero in it, in *Epigraphia Indica* 34 (1961–1962).
3. Moritz Cantor, *Vorlesungen uber Geschichte der Mathematik* vol. 1 (Leipzig: Druck & Teubner, 1891), 608.
4. Louis C. Karpinski, "The Hindu-Arabic Numerals," *Science* 35, no. 912 (June 21, 1912): 969–70.
5. Ibid., 969.
6. G. R. Kaye, "Indian Mathematics," *Isis* 2, No. 2 (September 1919): 326.
7. Ibid., 328.
8. I am grateful to Takao Hayashi for this information. He discusses the lost Khandela tablet in his book in Japanese, *Indo no sugaku* [Mathematics in India] (Tokyo: Chuo koron she, 1993), 28–29.

CHAPTER 10

1. A good modern source is Pich Keo, *Khmer Art in Stone*, 5th ed. (Phnom Penh: National Museum of Cambodia, 2004).
2. George Cœdès, "A propos de l'origine des chiffres arabes," *Bulletin of the School of Oriental Studies* (University of London) 6, no. 2 (1931).
3. Ibid.
4. Ibid., 328.

CHAPTER 11

1. Chou Ta-kuan, "Recollections of the Customs of Cambodia," translated into French by Paul Pelliot in *Bulletin de l'École Française d'Extrême-Orient*, 123, no. 1 (1902): 137–77. Reprinted in English in *The Great Chinese Travelers*, Jeannette Mirsky ed. (Chicago: University of Chicago Press, 1974), 204–6.
2. Ismail Kushkush, "A Trove of Relics in War-Torn Land," *International Herald Tribune*, April 2, 2013, 2.
3. C. K. Raju, "Probability in India," in *Philosophy of Statistics,* Dov Gabbay, Paul Thagard, and John Woods, eds. (San Diego: North Holland, 2011), 1176.
4. Nagarjuna, *The Fundamental Wisdom of the Middle Way* (Oxford, UK: Oxford University Press, 1995), 3.
5. Ibid., 73.
6. Thich Nhat Hanh, *The Heart of the Buddha's Teaching* (New York: Broadway, 1999), 146–48.

7. George Cœdès, "A propos de l'origine des chiffres arabes," *Bulletin of the School of Oriental Studies* (University of London) 6, no. 2 (1931) 323–28.

CHAPTER 16

1. This comes from Graham Priest, "The Logic of the *Catuskoti*," *Comparative Philosophy* 1, no. 2 (2010): 24.
2. T. Tillemans, "Is Buddhist Logic Non-Classical or Deviant," 1999, 189, quoted in Graham Priest, "the Logic of the *Catuskoti*," 24.
3. S. Rhadakrishnan and C. Moore, eds., *A Sourcebook on Indian Philosophy* (Princeton, NJ: Princeton University Press, 1957), quoted in Graham Priest, "The Logic of the *Catuskoti*," 25. Priest explains that "saint" is a poor translation and that what it means is someone who has reached enlightenment, a Buddha (or Tathagata).
4. Graham Priest, "The Logic of the *Catuskoti*," 28.

CHAPTER 17

1. For more on the story of Georg Cantor and the various levels of infinity, see Amir D. Aczel, *The Mystery of the Aleph* (New York: Washington Square Books, 2001).

CHAPTER 22

1. For accurate radiocarbon dating of the Thera explosion see Amir D. Aczel, "Improved Radiocarbon Age Estimation Using the Bootstrap," *Radiocarbon* 37, no. 3 (1995): 845–49.

CHAPTER 24

1. I heard this story from another well-known mathematician and friend of Kakutani, Janos Aczel (no relation; it's a common Hungarian last name).

CHAPTER 26

1. George Cœdès, "A propos de l'origine des chiffres arabes," *Bulletin of the School of Oriental Studies* (University of London) 6, no. 2 (1931): 326.
2. Ibid., 327.

CHAPTER 28

1. Some details about the house and its location have been changed to protect the occupants' privacy.

BIBLIOGRAPHY

Artioli, G., V. Nociti, and I. Angelini. "Gambling with Etruscan Dice: A Tale of Numbers and Letters." *Archaeometry* 53, no. 5 (October 2011): 1031–43.

Aubet, M. E. *The Phoenicians and the West.* Cambridge: Cambridge University Press, 2001.

Saint Augustine. *The City of God.* New York: The Modern Library, 2000.

Boyer, Carl B., and Uta Merrzbach. *A History of Mathematics.* 2nd ed. New York: Wiley, 1993. This is a standard scholarly source on Babylonian, Egyptian, Greek, and other early mathematics, including a description of the early Hindu numerals; it does not include the discoveries of the earliest zeros in Southeast Asia.

Briggs, Lawrence Palmer. "The Ancient Khmer Empire." *Transactions of the American Philosophical Society* (1951): 1–295. Information on some now-lost inscriptions with early numerals from Cambodia.

Cajori, Florian. *A History of Mathematical Notations.* Vols. 1 and 2. New York: Dover, 1993. A reissue of a superb source of information on mathematical notations; it does not include the discoveries of the earliest numerals in Southeast Asia.

Cantor, Moritz. *Vorlesungen uber Geschichte der Mathematik.* Vol. 1. Berlin, 1907.

Cœdès, George. "A propos de l'origine des chiffres arabes." *Bulletin of the School of Oriental Studies* (University of London) 6, no. 2 (1931): 323–28. This is the seminal paper by Cœdès, which changed the entire chronology of the evolution of our number system by reporting and analyzing the discovery, by Cœdès himself, of a Cambodian zero two centuries older than the accepted knowledge at that time.

Cœdès, George. *The Indianized States of Southeast Asia.* Hilo: University of Hawaii Press, 1996. A comprehensive, authoritative source on the history of Southeast Asia with references to the author's work on discovering the earliest numerals.

Cunningham, Alexander. "Four Reports Made During the Years 1862–1865." *Archaeological Survey of India* 2 (1871): 434.

Dehejia, Vidaya. *Early Buddhist Rock Temples.* Ithaca: Cornell University Press, 1972. An excellent description of Buddhist rock and cave inscriptions, including very early numerals.

Diller, Anthony. "New Zeros and Old Khmer." *Mon-Khmer Studies Journal* 25 (1996): 125–32. A recent source on early zeros in Cambodia dated to the seventh century.

Durham, John W. "The Introduction of 'Arabic' Numerals in European Accounting." *Accounting Historians Journal* 19 (December 1992): 25–55.

Emch, Gerard, et al., eds. *Contributions to the History of Indian Mathematics.* New Delhi: Hindustan Books, 2005.

Escofier, Jean-Pierre. *Galois Theory.* Translated by Leila Schneps. New York: Springer Verlag, 2001.

Gupta, R. C. "Who Invented the Zero?" *Ganita Bharati* 17 (1995): 45–61.

Hayashi, Takao. *The Bakhshali Manuscript: An Ancient Indian Mathematical Treatise.* Groningen: Egbert Forsten, 1995.

Hayashi, Takao. *Indo no sugaku* [Mathematics in India]. Tokyo: Chuo koron she, 1993.

Heath, Thomas. *A History of Greek Mathematics,* Vol. 1. New York: Dover, 1981.

Ifrah, Georges. *The Universal History of Numbers.* New York: Wiley, 2000. This is a well-recognized, comprehensive work on the history of numbers and is much quoted. It is, however, neither very scholarly nor based on original research. The fact that it receives continuing attention only points to the need for a very serious and deep analysis of this crucial step in humanity's intellectual history.

Jain, L. C. *The Tao of Jaina Sciences.* New Delhi: Arihant, 1992.

Kanigel, Robert. *The Man Who Knew Infinity: A Life of the Genius Ramanujan.* 5th ed. New York: Washington Square Press, 1991.

Kaplan, Robert, and Ellen Kaplan. *The Nothing that Is: A Natural History of Zero.* New York: Oxford University Press, 2000. A good source on the mathematical idea of zero, with some information on the development of the symbol, but not including the earliest appearances of this key symbol.

Karpinski, Louis C. "The Hindu-Arabic Numerals." *Science* 35, no. 912 (June 21, 1912): 969–70.

Kaye, G. R. "Notes on Indian Mathematics: Arithmetical Notation." JASB, 1907.

Kaye, G. R. "Indian Mathematics." *Isis* 2, no. 2 (September 1919): 326–56. Kaye's now-notorious manuscript discrediting Indian priority over the invention of numerals.

Keay, John. *India: A History.* New York: Grove Press, 2000. An excellent general history of India.

Keyser, Paul. "The Origin of the Latin Numerals from 1 to 1000." *American Journal of Archaeology* 92 (October 1988): 529–46.

Lal, Kanwon. *Immortal Khajuraho.* New York: Castle Books, 1967. A general description of the temples of Khajuraho.

Lansing, Stephen. "The Indianization of Bali." *Journal of Southeast Asian Studies* (1983): 409–21. Includes a description of number-related discoveries in Indonesia.

Mann, Charles C. *1491: New Revelations of the Americas Before Columbus.* New York: Knopf, 2005. Good description of the Mayan numerals and zero glyph.

McLeish, John. *The Story of Numbers.* New York: Fawcett Colombine, 1991.

Nagarjuna. *The Fundamental Wisdom of the Middle Way.* Translated by Jay L. Garfield. New York: Oxford University Press, 1995.

Neugebauer, Otto. *The Exact Sciences in Antiquity.* Princeton, NJ: Princeton University Press, 1952.

Nhat Hanh, Thich. *The Heart of the Buddha's Teaching.* New York: Broadway, 1999.

Nicholson, Louise. *India.* Washington, DC: National Geographic, 2014.

Pich Keo. *Khmer Art in Stone.* 5th ed. Phnom Penh: National Museum of Cambodia, 2004.

Plofker, Kim. *Mathematics in India.* Princeton, NJ: Princeton University Press, 2009. An excellent comprehensive source on the general developments in mathematics in India since antiquity.

Priest, Graham. "The Logic of the *Catuskoti.*" *Comparative Philosophy* 1, no. 2 (2010): 24–54.

Raju, C. K. "Probability in India." In *Philosophy of Statistics,* edited by Dov Gabbay, Paul Thagard, and John Woods, 1175–95. San Diego: North Holland, 2011.

Robson, Eleanor. "Neither Sherlock Holmes nor Babylon: A Reassessment of Plimpton 322." *Historia Mathematica* 28 (2001): 167–206.

Robson, Eleanor. "Words and Pictures: New Light on Plimpton 322." *Journal of the American Mathematical Association* 109 (February 2002): 105–20.

Smith, David Eugene. *History of Mathematics, volume 2: Special Topics in Elementary Mathematics.* Boston: Ginn and Company, 1925.

Smith, David Eugene, and Louis Charles Karpinski. *The Hindu-Arabic Numerals.* Boston: Gin and Company, 1911.

Ta-kuan, Chou. "Recollections of the Customs of Cambodia." Translated into French by Paul Pelliot, in *Bulletin de l'École Française d'Extrême-Orient,* No. 1 (123), (1902): 137–77. Reprinted in English in Mirsky,

Jeannette, ed. *The Great Chinese Travelers.* Chicago: University of Chicago Press, 1974.

Tillemans, T. "Is Buddhist Logic Non-Classical or Deviant?" In *Scripture, Logic, Language: Essays on Dharmakirti and his Tibetan Successors.* Boston: Wisdom Publications, 1999.

Wolters, O. W. "North-West Cambodia in the Seventh Century." *Bulletin of the School of Oriental and African Studies* (University of London) 37, no. 2 (1974): 355–84.

Zegarelli, Mark. *Logic for Dummies.* New York: Wiley, 2007.

INDEX